Plantations and Protected Areas

History for a Sustainable Future

Michael Egan, series editor

Derek Wall, *Culture, Conflict, and Ecology: The Commons in History*

Frank Uekötter, *The Greenest Nation? A New History of German Environmentalism*

Brett M. Bennett, *Plantations and Protected Areas: A Global History of Forest Management*

Plantations and Protected Areas

A Global History of Forest Management

Brett M. Bennett

The MIT Press
Cambridge, Massachusetts
London, England

MIT Press books may be purchased at special quantity discounts for business or sales promotional use. For information, please email special_sales@mitpress.mit.edu.

This book was set in Sabon LT Std by Toppan Best-set Premedia Limited. Printed and bound in the United States of America.

Library of Congress Cataloging-in-Publication Data

Names: Bennett, Brett M., 1983–
Title: Plantations and protected areas : a global history of forest management / Brett M. Bennett.
Other titles: Global history of forest management
Description: Cambridge, MA: The MIT Press, 2016. | Series: History for a sustainable future | Includes bibliographical references and index.
Identifiers: LCCN 2015037897 | ISBN 9780262029933 (hardcover: alk. paper)
Subjects: LCSH: Forest protection. | Sustainable forestry. | Forest reserves.
Classification: LCC SD411 .B46 2016 | DDC 634.9/2—dc23 LC record available at http://lccn.loc.gov/2015037897

10 9 8 7 6 5 4 3 2 1

Contents

Series Foreword

Michael Egan

As the crickets' soft autumn hum
is to us,
so are we to the trees

as are they

to the rocks and the hills.[1]

On a recent visit to Oxford, I happened upon a haunting metaphor of global forests and their future. It was a wet spring and the Cherwell River had flooded. My carry-on travel bags did not permit me to bring a variety of footwear options, so I didn't have with me the necessary shoes for tromping across the soggy fields that separate the town and colleges from Marston, where I was staying with family. The Romans might have marched straight across the fields, but they would have been better shod. So I stuck to the pavements, taking the more circuitous route toward Summertown and then down the Marston Ferry Road. The route took me past the Museum of Natural History, which had on its front lawn a rather striking exhibit called "Ghost Forest," a collection of massive tropical forest tree stumps, mounted—lying down—on stone slabs.[2] I tarried. It was late afternoon, and as I wandered from installation to installation, reading each plaque—which indicated the tree species and its full, living height in a manner that felt more like eulogy than informative display—the world became rather quiet. The traffic and bustle of the road, not twenty feet away on the other side of the waist-high stone wall, was muted. I was in a mausoleum, or the silent aura one associates with entering a church.

The exhibit was surprisingly moving: less, perhaps, the explicit ghost forest message, and more, simply, the massive remnants of once-living things. Shelley might have written a more sympathetic version of "Ozymandias" for them, where just the stumps remain. I was transported back to childhood holidays on Vancouver Island and trips to Cathedral Grove. There, Douglas firs older than the Oxford colleges, through which I had roamed prior to discovering the ghost forest, still stand protected from the axe, accessible to visitors. Equally massive, awe-inspiring. Or the Carmanah Valley, further west on North America's edge. During my early adulthood, the struggle to protect the Carmanah resonated like few other political debates for me. On the one hand, the fight over the Carmanah Valley was one of many similar efforts to preserve the unique, natural beauty of old-growth forests and to demonstrate the social and economic value of trees uncut. On the other, it was an attempt to highlight the precarious fragility of ancient ecosystems and to improve our ethical relationship with some abstract nature. Though close to home, I had never visited the Carmanah. But, still, they elicited a sense of home, of belonging, of importance.

A significant portion of environmental history's mission is to highlight human trespasses into nature. Themes of resource extraction, landscape despoliation, scarcity, and sustainability abound in the literature. Global and transnational analyses of these topics also investigate the social, economic, and historical factors that explain the more rapid rate of deforestation in the tropical world in relation to the increased protection of old-growth stands in the northern, more prosperous parts of the world (a far from simple, perfect, or complete distinction). The ghost forest was a warning—a testament. But in Oxford, I was especially struck by the sadness I felt. Maybe it was the fatigue of travel catching up with me, combined with mounting homesickness. But the sadness seemed to be born of a kind of kinship—very distant cousins, as it were. It felt more like a palpable reminder of the larger community of life, to which we are all connected.

Deforestation is another expression of the Great Acceleration. Whereas tropical forests—measured by acreage or amount of old growth—have been declining for centuries, that process gained speed with European expansion and increased at a colossal rate after World War II.[3] The demand for timber, but also for the land underneath it to be converted to ranching and agriculture, has radically transformed ever-growing swaths of the tropical world. In turn, that transformation has altered and moved peoples, economies, and politics, not to mention their relationships with power and the environment. Timber scarcity and responses to it are not novel refrains, though. Indeed, the parallel narrative to forest degradations is the emergence of a scientific forest management designed to remedy the potential for economic and ecological collapse.

Trying to procure more trees with less land is at the heart of Brett Bennett's treatise on global forestry and forest management. Sustaining or increasing timber resources on diminishing lands, while ensuring that ecosystems remain healthy, is a daunting task, one that has historically been littered with "conspiracies of optimism," abject failures, and hopeful successes.[4] The expansive narrative that Bennett presents echoes the dominant faith in Western science that shaped Diana Davis's work on French colonialism and agriculture in northern Africa.[5] But this is an older story than the science of forest management that emerged in the wake of industrial scarcity. Paul Warde and others have shown how silviculture's prehistory stems from much earlier, local worries about imminent scarcity.[6] In its first iterations, "sustainability" was a concept derived to best manage dwindling timber resources in parts of Germany and, later, in England, where John Evelyn stressed the significance of healthy and abundant forests to Britain's naval survival.

From the local to the global: forest resources didn't seem so scarce. North American Eden narratives—a continent of infinite resources—and European colonization of Asia meant that trees were perhaps further away, but no less abundant. Still, early colonial expansion resulted in some of the most prescient efforts to conserve resources. The establishment of forest reserves from Europe to Mauritius to Indonesia acknowledged that uncontrolled, rapacious plunder had deleterious consequences.[7] An ordering of landscapes evolved into a system of management, which created a global agricultural system that treated trees as crops. Forest plantations were distinguished from protected areas. Exotic species could be introduced and grown and cultivated in structured patterns. We might think of the forest plantations as sacrifice zones, adopted to allow for the protection of native species.[8] Managing these distinctions, while nurturing both types of forests, is the enduring challenge of foresters the world over.

Another part of the juxtaposition between trees for cutting and trees for saving is that Bennett's protected areas are far more than trees that have been spared the axe. In more contemporary parlance, these protected areas are the lifeblood of biodiversity preservation and the centerpiece of conservation biology. Put another way, native forests are the stubborn holdfasts against exotic and invasive species, even if such strict distinctions are never so tidy. Emerging after World War II, during the Age of Ecology, Bennett notes that these protected areas and forests reserves—many of them established by colonial foresters—have become key battlegrounds. But they also constitute an abrogation of forester authority. On the ground, in the courts, and in legislative houses, the efforts to protect forests, trees, and their inhabitants have diminished forest managers' agency (in some cases, maybe, for the better). In effect, the shift to insert "broader social and environmental considerations"—as distinct from century-old commercial priorities—into forest debates has altered the forester's role.[9] And this introduces some interesting questions about the changing role of scientific expertise in environmental issues.

This is the morass into which Bennett wades. And it points to the critical importance of history in making sense of the world in which we live. Examining catastrophe and its social implications, sociologist Robert Wuthnow observes that we are not very good at responding to threats—terror, pandemic, or environmental. A big part of this, he concludes, stems from our need to act, but we typically respond like generals fighting the last war: Terror provokes Cold War reactions; avian flu elicits revisiting the 1918 epidemic as a reference point. "A threat appears and we buy duct tape," he writes, "because we did that the last time."[10] This is Wuthnow's problem with history: antecedents dictate actions even when they are not especially useful. Bennett makes a related case for our current global forest crisis. Contemporary forest management and the struggle to reduce wasteful deforestation worldwide is fraught with twentieth-century approaches and understandings (as varied and complex as those were), and burdened by dominant-but-outdated silvicultural practices and policies.

Wuthnow's problem with history does not negate history's value, however. Rather, it magnifies its importance. If history is simply a matter of recalling events, then we are indeed doomed to repeat it as Wuthnow describes. But history is the study of nuance and complexity. It is the process of disentangling and demystifying the myriad events, decisions, and knowledges that have shaped our present. Identifying how continuity and change have evolved through global forestry's past provides a potent lens for reading the present landscape and planning for a more resilient future. In charting such an ambitious history of past forests, Bennett provides a roadmap for imagining strategies to realize robust future forests. Perhaps it is a foolish question, but what will forests look like in the future? Bennett shows we are at a critical tipping point where globalization's contemporary economic pressures are in conflict with long-term resilience imperatives.

Acknowledgments

Many friends, colleagues, and family members have encouraged me during the writing of this book. I am thankful for all of your support. I am unfortunately unable to individually thank all of the people who offered encouragement and advice at various stages of the project owing to word limitations.

A few people who played a special role in helping this book come to print deserve mention. Fredrik Albritton-Jonsson, Diana Davis, Simon Pooley, and Gregory Barton inspired me to start and finish the project. Frederick Kruger commented on the manuscript in various incarnations. John Dargavel and James Beattie advised on specific chapters. Michael Egan, the series editor, has encouraged me to undertake and finish this book since we first talked in 2010. Beth Clevenger, the MIT Press acquisitions editor, has shown great generosity and professionalism in helping me revise the manuscript for publication. The MIT Press team, especially Kathleen Caruso, has helped me at every step of the publishing process.

The Forest History Society (special thanks to Cheryl Oakes, Steven Anderson, Jamie Lewis, and Eben Lehman) has given me consistent financial and scholarly support, including providing most of the pictures in this book. My colleagues at the University of Western Sydney (especially Peter Hutchings, Tim Winter, Tim Rowse, Robert Lee, Ien Ang, Lynette Sheridan

Burns, and Paul James) have inspired me since I joined the faculty in early 2011. Special thanks go to my mentors (Bruce Hunt, Wm. Roger Louis, Tony Hopkins, Gail Minault, James Vaughn) and fellow graduate students at the University of Texas at Austin. Faculty at the University of Cape Town (Lance van Sittert), Stellenbosch University (David Richardson and Sandra Swart), and the Australian National University (Tom Griffiths, Carolyn Strange, Libby Robin, Melanie Nolan, Kim Sterelny, Peter Kanowski, and all the members of the Centre for Environmental History and the School of History) have offered me an intellectual home away from home for many years.

The University of Western Sydney, Australian Research Council (DP140102991), National Science Foundation (USA), and Social Science Research Council (USA) funded my research.

Introduction: The Forest Management Divergence

A myriad of human-caused pressures threaten the world's forests. Demand for wood products and agricultural land endangers many of the last unlogged tropical forests. Tropical forests were deforested from 2000 to 2010 at a yearly rate of 5.2 million hectares, or roughly the size of Costa Rica, to make way for farms, plantations, and ranches. Trade liberalization since the 1980s has allowed Western corporations to export timber processing jobs and to import wood fiber and forest products from developing countries with lower costs and ideal growing conditions. Exotic species now comprise a sizeable percentage of the forest cover in developing countries in Asia, Africa, and South America. Ecologists warn that this is increasing the potential that exotic trees will become invasive in the future. These are but a few of the most pressing problems threatening the world's forests.

As a result of these changes, some scientific experts warn that the world's forests face historically unprecedented crises. In 2003, the eminent forest ecologist Jerry Franklin urged participants in forest policy discussions to look to the future, rather than the past, to find the solutions needed to address contemporary problems. He argued:

> Too many participants in the current forest policy debates—stakeholders, media, politicians, and resource professionals—appear

focused on the past rather than on the future . . . In North America and many other regions of the world, stakeholders and politicians continue to fight the resource war of the 20th century—preservation versus exploitation. These battles are familiar and comfortable. But the major challenges of the 21st century are not likely to repeat those from the previous century. Few of the forest policy debates, alliances, and "solutions" of the 20th century are likely to be relevant to these new challenges.[1]

Franklin noted that economic globalization was the primary factor reshaping forest policy throughout the world: the "globalisation of the wood products industry is, I believe, the most significant factor influencing the developing context for forest stewardship."[2]

Franklin's assertion that many participants have focused narrowly on conflicts over the management of public forests without understanding broader global change is correct, although we should be wary of the claim that the past cannot help us explain and solve contemporary forestry problems. From a historical perspective, many of the world's "new" environmental problems—rapid deforestation, climate change, pollution, and peak resource use—appeared during the modern era (1750–present).[3] Nor are the processes associated with globalization new: we are living through the most recent phase of a longer process of global integration and connection that began in earnest during the early modern era (1500–1750) and continues to today.[4]

Almost all of the management paradigms and problems that scientific experts and policy makers now wrestle with appeared in their current forms by the 1980s. The expansion of intensive exotic timber plantations in tropical countries, the creation of protected forest areas, the privatization of state plantations, fears about tropical deforestation, and the awareness that trees can be problematic invasive species garnered considerable attention during the 1980s. If we look even further back, the antecedents to these trends can be traced back to the eighteenth and nineteenth centuries.

The historical perspective offered in this book demonstrates that there has been a divergence of wood production from the protection of forests across the twentieth century that is being caused by entangled environmental, social, political, and economic processes that first appeared in the eighteenth and nineteenth centuries. I describe this trend as the *forest management divergence*. The two key processes driving this forest management divergence—an intensification of timber production through plantations alongside the concomitant creation of protected areas—are reshaping other forest management systems, including privately owned forests and forests that are under communal and indigenous tenures. The declining cost of wood fiber and the increasing value of biodiversity and carbon sequestration impacts how people throughout the world manage trees and forests.

The wood fiber and timber products that we consume now come primarily from intensive plantations or seminatural forests. *Seminatural forests* are defined as forests that have been significantly modified by human activity and are managed for production and conservation purposes. These forests include large coniferous forests in northern Europe (including parts of Russia), the South and the Pacific Northwest in the United States, and western and eastern Canada. Most of these forests have been managed for timber production as well as environmental conservation. A growing proportion of wood comes from *plantations*, forests created specifically for commercial timber production. Most new plantations are *intensive* plantations composed of exotic species that have been selected and bred to produce desirable timber qualities and grow quickly. Today, over 35 percent of industrial wood is supplied from intensive plantations. This figure is predicted to reach 50 percent by 2020.

In response to the growing human impact on the environment, scientists and concerned citizens are pushing governments to lock up the world's remaining *native forests*, forests that have had less modification because of human activity, in *protected areas*. Protected areas are managed based on a series of guidelines laid down by the International Union for Conservation of Nature, the UN Convention on Biodiversity, and the World Heritage Convention. Governments that manage protected areas are required by these guidelines to limit commercial extraction in an attempt to preserve biodiversity. The justification for protected areas was originally to preserve biodiversity, but there has been a push to protect forests for their role in sequestering carbon. Current scientific research indicates that there is twice as much carbon locked in the world's trees as there is in the atmosphere.

Paradoxically, plantations and protected areas are management models that, in theory, should complement each other, although in practice they often contradict. In theory, plantation-grown timber can alleviate the need to harvest from native forests, thus allowing for the creation of more protected areas. Yet in practice advocates of plantations and protected areas are odd bedfellows because they are at the two extremes of forest policy—extraction and preservation. There have been hopeful signs that agreements between the timber industry and environmentalists can accommodate both protected areas and plantations in a single scheme. But more often than not these schemes have been abandoned or less successful than desired because of antagonism between proponents of protected areas and those who seek to expand harvesting in public forests.

Various reasons account for these failures, but much of the antagonism felt in current policy debates can be attributed to the fact that most countries lack a credible arbiter who can help determine and implement forest policy. Professional foresters working on behalf of governments fulfilled such a mediating role for most of the twentieth century, but the loss of public and scientific confidence and the economic pressures of globalization since the 1980s have eroded their influence. States remain the most important player in directing forest policy discussions because 80 percent of the world's forests are government controlled. Yet instead of providing active leadership, many governments have allowed free-market forces to direct the production of timber while doing relatively little to protect forests located in designated protected areas. This new bifurcated policy directly contradicts the older model of forest management, the conservation model, from which plantations and protected areas emerged. To understand the bifurcation of forest policy into distinct protective and productive models, we must understand the longer history of forest management.

The Conservation Model

Plantations and protected areas both evolved out of an older integrated forest management system that I describe in this book as the *conservation model* because this model sought, in the broadest sense, both to produce timber resources and to conserve the environment. The state implemented the conservation model in response to concerns about timber scarcity and fears about the negative consequences of deforestation. The conservation model relied on the expertise of professionally trained scientific foresters to make decisions about how to manage public forests within the territorial boundaries of a state. Foresters were given power by governments to manage forests and to mediate between the interests of private industry and those of the public.

The conservation model spread throughout the world in the second half of the nineteenth century in response to a dynamic phase of modern globalization. Almost every government in the world created forestry legislation and bureaucracies with the same overarching premises:

- Governments should control forests as a public good.
- Professional foresters should manage government-controlled forests.
- Forest management policies would both sustainably extract timber while also maintaining forest cover as a means to conserve soil, water, and even climate.
- Governments would in *exceptional* circumstances create national parks to preserve superlative sites of natural beauty and naturalistic interest. Foresters often played an important role in determining early national parks management policies.

These definitions describe the legal and bureaucratic structure of the majority of forest and conservation programs worldwide until the 1980s.

The dual goals of timber production and forest conservation began to fragment in divergent directions during the twentieth century. Timber plantations emerged as a viable way to produce timber at the same time that alternative forms of environmental protection, such as national parks and wilderness areas, undermined the legitimacy of the conservation model. These two models—plantations and protected areas—started to undermine the conservation model in the 1980s. Governments were influenced by social protests, growing environmental problems, and the financial burden of supporting large state forestry bureaucracies. Governments in countries as diverse as New Zealand, China, Thailand, the United States of America, and Australia all lowered the harvest rate in state forests when they created new protected areas for forests. China now has around 90 million hectares of protected forests and almost all of its domestic timber production comes from exotic intensive plantations. Governments in New Zealand,

Australia, the United Kingdom, and South Africa privatized or leased state-owned plantations in order to offload timber plantation onto the private sector. The shift toward plantation production gained pace in the 1990s and 2000s as a result of trade liberalization and the rapid economic growth of developing countries.

Local Constraints, Global History

The history of modern forest management can be conceptualized as an interaction between human attempts to regulate forests and the constraints that shaped the outcome of these efforts. Material environmental factors—biological, ecological, climatic, and geographic conditions—have limited the degree to which humans have been able to change environments. Humans have been able to study and manipulate environmental constraints—for example, through irrigation or selective breeding—but they have always been limited to some degree in their ability to control individual species or the wider environment. Social constraints, such as political movements to control harvesting in government-controlled forests, can stop the implementation of management policies. Economic factors, such as access to markets and capital, constrain management decisions. Limitations in technology make it difficult to utilize certain timbers, making them less economically valuable. All of these constraints have worked together to shape policy outcomes.

Misconceptions can arise if we view forest management in purely political or intellectual terms without acknowledging factors that constrained outcomes. Many studies of forestry have suggested that forestry policies throughout the world followed a similar model of forestry that originated in Germany and diffused throughout the world in the nineteenth century. German forestry is portrayed as a management program that sought to turn, and succeeded in turning, diverse native forests into a "one commodity machine" with the single-minded goal of producing timber.[5] James Scott argued that this German model became "standard throughout the world."[6] Other scholars have agreed that German forestry methods were transferred without significant modification to other locations throughout the world.[7]

The idea that professional foresters reshaped the world's diverse forests into the image of a simplified German spruce plantation has been challenged in recent years by scholars examining the application of forest management policies in other locations around the world. The historical anthropologist K. Sivaramakrisnan argued in a somewhat exasperated tone in his influential study of forestry in eastern Bengal: "I am forced, by recent scholarship in very different genres, to insist on this point . . . scholars consistently conflate policy intent with practical outcome."[8] What he meant is that foresters, in Bengal and elsewhere, generally lacked the capability to reshape forests to attain economic or technical ideals. Even if foresters did want to engineer nature purely for production, a myriad of constraints limited their efforts to do so. Sivaramakrisnan emphasized the need to take into account local processes and the unique history of regions when writing about the history of forest management in order to capture the complexity of forest history.[9] His view echoes similar writings in the field of historical geography. David Livingstone called on historians to "put science in its place" by carefully examining how local conditions reshaped seemingly "universal" scientific principles and knowledge.[10]

These methods are particularly useful for conceiving of the forest management divergence, a process that is inextricably bound up with modern globalization. Globalization has been a complex process characterized by multiple scales of causation that included local microprocesses, from individual action to nonhuman actors to global trends, such as international trade or human or nonhuman migration. Globalization in the modern era cannot be characterized as a simple one-way process of diffusion from Europe outward. Global movements of people, plants, and ideas were mediated by adaptation, resistance, rejection, and counterflows. Non-European precedents and problems played a significant role in shaping dominant forest management assumptions, such as the belief that protecting forests ameliorates the climate. Many ideas about nature that emerged from European conditions failed to explain environments in other parts of the world, especially the tropics.

Global history raises significant challenges, not the least being that it is difficult to write a global history without making some generalizations and choosing selectively to emphasize key trends and ideas. Global trends, those that impact on multiple sites, always do so *unevenly*. For instance, the movement to "lock up" forests is most apparent in postcolonial Anglo settler colonies, South Africa, Australia, New Zealand, and the United States where settlers alienated indigenous people from iconic landscapes and reimagined these spaces as "wilderness." This trend has not been as relevant in Europe, where ideas of "wilderness" and "national parks" were less historically important because higher population densities and longer human habitation have significantly modified landscapes. So too can one identify the expansion of exotic timber plantations in some developing countries, such as Indonesia and Brazil, but not in all. The examples used in this book are meant to highlight key global trends, but it must be acknowledged that given word limitations and the expertise of the author certain countries and examples receive more attention than others. More attention is directed to developed countries because developing countries have only become drivers of global forest policy in the final decades of the twentieth century.

Defining Forests

Forests currently clothe one-third of the world's landmass, or approximately four billion hectares in total. A satellite picture of earth from outer space reveals two major bands of forest-cover located in the tropics between the Tropic of Cancer and the Tropic of Capricorn and between the 40th to 70th parallels in the Northern Hemisphere. Most of the Southern Hemisphere (outside the tropics) is dominated by desert, grassland, savanna, and a small fringe of forests hugging the coasts and mountains of South America, southern Africa, New Zealand, and Australia. As I discuss in the book, these broad biogeographic patterns have played a fundamental role in the evolution of forest management practice and theory from the eighteenth century until the present day.

This book uses the term "forests" to describe a grouping of trees that is managed for either production or conservation. The definition encompasses the vast majority of the world's forests, 80 percent of which are claimed by governments. Almost all government-controlled forests are by law designated for conservation or production. These government forests are directed through a variety of laws and management programs. In most forests, governments strictly regulate the utilization of forest products. There are communal and indigenous tenure schemes, which are prominent in many developing countries, that give local people greater power to determine the utilization and management of forests. Around 20 percent of the world's forests are located on privately owned land. These forests are used for a variety of purposes, including mixed production and conservation, industrial timber production, or biodiversity preservation. Privately owned forests account for over 50 percent of forests in the developed countries of Sweden and the United States.

Forests can be classified using other definitions. Definitions can be made based on ecological succession (old growth, secondary growth, etc.), harvesting regime (clear-felling, selective felling, etc.), management model (intensive plantations, shelter belts, protected areas), and ownership (private, public, communal), among other things. Most scientific definitions of forests focus on naturalness, the extent to which human action has purposefully or accidentally shaped the ecological composition and life history of a forest. There is a growing consensus that all of the world's forests, even the most remote rain forest, have been shaped by human action, even though it has often been controversial to make this point.[11] This does not mean that a remote Brazilian rain forest is equivalent to a plantation of exotic *Eucalyptus* in the eastern part of the country. But we should be careful not to see forests as places without people. Humans and nature are inextricably bound together now as a result of historical and contemporary impacts on the environment.

Despite pitfalls, there is value in defining forest types based on their environmental condition. This book draws distinctions between forest types—"plantation," "seminatural forest," and "native forest"—to explain how management models have reshaped ecological systems. It is, of course, easy to define forests that are clearly an extreme example, such as a commercial intensive plantation or an old-growth protected forest. Marking where definitional boundaries begin and end is somewhat trickier. For instance, is a two-hundred-year-old spruce-dominated forest located in northern Europe a plantation because it was initially created in the nineteenth century to produce timber, although people later valued the forest for its protective functions? Or is it a native forest because it is composed of species native to the region where it is located? Maybe it is a seminatural forest because it has diversified ecologically

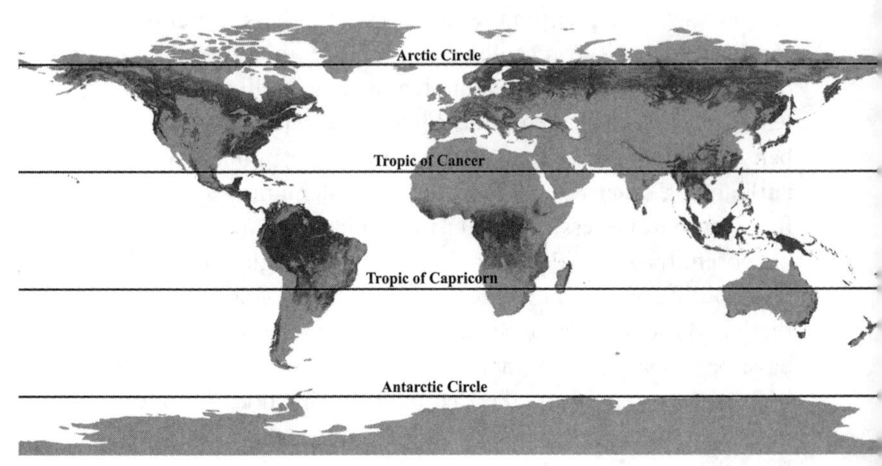

Figure 0.1

Global map of forest cover, 2014

Source: M. C. Hansen et al., "High-Resolution Global Maps of 21st-Century Forest Cover Change," *Science* 342, no. 6160 (2013): 850.

since its creation and has been a feature of the landscape for centuries?

The definitions used in the book serve the purpose of helping to clarify the historical argument rather than making any moral or scientific statement about different forest types globally. The book seeks to acknowledge the various purposes of forests in the past and present in order to accommodate a variety of views about forests. For instance, this book defines the large forests of northern and central Europe as seminatural forests because they are primarily composed of native species and they serve a variety of purposes, including production and environmental protection. The definition acknowledges the various ideas and management plans that shaped many of Europe's forests without making a judgment about which model is best.

Outline

Chapter 1 traces the evolution of the conservation model from its early modern origins until the middle of the twentieth century. In the late 1400s, European states started to create forest laws and management programs to regulate the harvesting of timber to promote security and economic stability. The science and profession of forestry developed in Europe in response to the emergence of professional science and environmental expertise during the late eighteenth and early nineteenth centuries. In the first half of the nineteenth century, naturalists in Europe and its overseas colonies drew on examples from India and tropical islands to argue that deforestation led to declining rainfall, extreme temperatures, and erosion. Governments throughout the world passed forestry legislation and founded forest services because they believed that conserving forests for production and conservation was a public good. Throughout the world, the development of forest management practices developed according to local environmental, social, and political conditions that limited the applicability of European methods. In particular, people living in tropical regions or those devoid of trees found it difficult to apply European methods.

Chapter 2 focuses on the evolution of management methods for establishing timber plantations. The history of timber plantations is a subject that has received surprisingly little research. Plantations became an important management model that foresters used to regulate forests in Europe. Outside Europe, plantations were usually created only after native forests were destroyed or when they could not provide useful timbers. Settlers in the United States, Australia, and Brazil depleted their native forest resources before creating timber plantations during the first half of the twentieth century. Governments in timber-deficient regions invested in the creation of plantations and funded research on plantations that made them economically viable. The creation of international development aid programs and national efforts to establish plantations in tropical climates caused an expansion of plantations in developing countries in the 1960s and 1970s. Economic changes in the 1980s and 1990s allowed developing countries to become significant producers of wood fiber used for paper and pulp.

Chapter 3 examines how ideas about forest management changed from the nineteenth to the late twentieth centuries in response to the expanding idea of "naturalness." In the nineteenth century, foresters dominated scientific research on forest systems. Some foresters began to criticize forestry management schemes because of their aesthetic and biological consequences. These criticisms expanded in the first half of the twentieth century as researchers in the fields of hydrology, wildlife biology, and ecology undermined key assumptions about forest conservation. The expansion of clear-felling as a harvesting method after World War II led to public conflicts over the management of state-controlled forests in countries such as the United States, Canada, New Zealand, Australia, and India. Deforestation caused by economic growth in developing countries created controversy as well. In response to social pressure, governments in many countries began to set aside large native forest areas as protected areas that would preserve the biological and ecological integrity of forests. The decline of harvesting from protected areas was more than offset by the increase in plantation-grown timber domestically and globally.

Chapter 4, the concluding chapter, discusses how the bifurcation of forest management policies is reshaping the production of timber and the protection of native forests. Some experts have called for the creation of national policies for locking away the majority of native forests in protected areas while producing industrial timber almost exclusively from commercial intensive plantations. Yet there are problems associated with the decoupling of production from protection. The creation of protected areas has not been followed by an increase in funding to actively manage ecological fragmentation, climate change, and invasive species. Intensive plantations have significant ecological and social impacts on the regions where they have been planted. The continuation of forest wars has led to a loss of knowledge of forestry that may be necessary to manage large ecosystems for diverse purposes. The book concludes by suggesting that a middle path may be more preferable than a total bifurcation.

1

The Conservation Model: Universal Pattern, Local Adaptation

"Remember—only YOU can prevent forest fires!" With these words, the cartoon character Smokey the Bear warned Americans in 1947 to be vigilant when they visited the country's vast national forest system. Smokey the Bear first appeared in advertisements toward the end of World War II when the Wartime Advertising Council conceived of him as a tool to encourage Americans to protect the country's forest resources, then in high demand for the war effort. After the war ended, Americans came to accept Smokey as an iconic part of American culture. Smokey the Bear continues to warn Americans in the twenty-first century that careless acts can cause forest fires.

Smokey the Bear encapsulated the three key values that shaped the management of federally controlled forestland in the United States throughout the twentieth century: extraction, conservation, and recreation. Protecting against fire ensured that U.S. federal forests could provide a sustainable yield of timber for future generations. Forests also housed a wide variety of plants and animals, such as bears, that required expansive forest ecosystems to survive. Making a real Smokey the Bear made this link clear to the American people. When firefighters found a bear cub hidden inside the cavity of a burnt log in the Lincoln National Park in New Mexico after a forest fire in 1950, the national media named him "Smokey," and he went off to live the rest of his life in the National Zoo in Washington, D.C., as a living expression of why Americans should prevent forest fires. National advertisement campaigns featuring Smokey encouraged Americans to visit the country's system of national forests and parks for recreation.

There is nothing particularly exceptional about American forest history—the use of cartoon animals to warn about fire or the threefold emphasis on extraction, conservation, and recreation—when we situate it within the context of global history. Other countries have employed iconic species to warn about the dangers of fire. In South Africa, the Cape Department of Nature Conservation used an antelope, Bokkie the Grysbok, to warn about the dangers of human-caused fire. Forestry programs elsewhere sought to protect iconic species. Indian legislation has afforded wild elephants protected status in state forests since the passing of The Elephants Preservation Act of 1879. National forests became sites of middle-class tourism in almost every white settler colony. Seven years after the United States created Yellowstone National Park in 1872, the parliament of New South Wales (Australia) created the world's second oldest national park in 1879 when it declared bushland to the south of Sydney as the National Park (later renamed Royal National Park) to give city residents a site for recreation. These examples do not detract from the importance of national developments, but they do tell us something important about the global history of forest management. They suggest that common global processes have shaped convergent forest management policies in the modern era, although in practice these policies varied according to local conditions.

The Conservation Model in Global Context

Smokey the Bear at first might seem far removed from the processes associated with globalization. Historians studying

Figure 1.1

The "real" Smokey Bear sees the cartoon Smokey at a promotional event, 1950

Credit: U.S. Forest Service photo courtesy of the Forest History Society, Durham, N.C.

globalization have demonstrated how the intensification of global connections during the past five hundred years has created patterns of convergence in different parts of the world. The use of cartoon animals like Smokey and Bokkie are examples of convergences in how states used public campaigns to warn against fire. Yet historians are also careful now to point out that these convergences varied greatly in terms of actual practices. So we should not read too deeply into superficial commonalities without examining how seemingly similar practices are played out on the ground. The challenge of writing global history is to connect how global processes that shaped multiple regions of the world influenced and were shaped by events and actions in specific localities.[1]

Forestry has evolved from the mid-eighteenth century until the present day. By "forestry" I refer to four interrelated, overlapping spheres: state policies, scientific inquiry, management practices, and professional norms. It is important to distinguish among these four spheres because the history of forestry has suffered from inadequate definitional clarity and a lack of understanding of the differences between forestry theories and actual practices.[2] Trying to assign a single "origin" to the conservation model is difficult because these aspects did not always develop at the same time or from the same location. Rather than finding a single putative origin, such as Germany, the prevailing model of modern forestry emerged from multiple sites in response to changes caused within intellectual and material life as a result of globalization.

States first managed forests to maintain domestic security against internal and external threats. The desire to maintain the output of forests in the face of possible shortages prompted European states to pass forestry legislation in the early modern period. Fears that timber scarcity could undermine state security continued to shape state forest policy globally into the twentieth century. States used anxieties about forest depletion to extinguish or limit "common rights" to forest by claiming control over forest resources.[3] The advent of scientific forestry (discussed below) gave states new justification for taking control of forests for wood security as part of the "territorialization" of forests, but the expansion of state power over forests began before scientific forestry emerged.[4]

State policies, scientific inquiry, management practices, and professional norms aligned in the first half of the nineteenth century when scientific forestry became a coherent discipline in Europe. The last half of the eighteenth century saw the emergence of professional environmental expertise throughout western Europe as governments gave patronage to naturalists to solve problems relating to the "organic economy."[5] Historians writing about forestry have tended to credit German foresters with having created the science and the profession of forestry. German foresters feature so prominently in histories focused on this period because they worked closely with the state to justify the hiring of forestry experts to use scientific methods to manage forests. While it is true that foresters dealt with unique problems associated with the long-term management of forests, they partook in the same debates and faced problems similar to those of naturalists studying other aspects of nature. Finding sustainability in the face of resource limitations was a problem that was felt throughout Europe. This chapter questions the story of German exceptionalism that pervades much of the literature of forestry.

The question of whether the state should intervene in the management of nature for the public good became a significant intellectual question in Europe and Europe's overseas colonies during the second half of the eighteenth century. Fredrik Albritton-Jonsson argues that Enlightenment resource management debates centered around two "rival ecologies of commerce" that pitted supporters of environmental regulation against free-market advocates who reasoned that market demand and the abundance of natural resources would inevitably sort out any imbalance in supply.[6] Throughout the nineteenth century, foresters argued that the private sector and individuals living near forests could not be trusted to sustainably manage forests in perpetuity. Well-known examples of deforestation in the United States, India, Brazil, and Australia during the nineteenth century helped convince governments about the need to regulate private interests. By the early twentieth century, governments espoused the principle that forest management was a public good, despite considerable resistance from business interests and local people living near forests. Gifford Pinchot, the first chief of the U.S. Forest Service (USFS), told the American public that the purpose of state forestry was to manage forests, "for the greatest good for the greatest number in the long run."[7]

A consistent framework of forest management shaped policies globally throughout the first three-quarters of the twentieth century. State forestry programs focused on two goals that at first seemed compatible but led to conflict later in the twentieth century. First, states implemented forestry programs to ensure the perpetual production of timber required by the economy and for the security of states. Foresters used a variety of management models, including plantations and mixed forests, to produce a wide variety of timber products. Foresters in Europe were fortunate because they had access to a number of species, the most valuable being the Norway spruce (*Picea abies*), that could be grown both in monoculture plantations and in mixed-forest management systems. The biology of native trees and the climatic conditions of northern and central Europe made it easier for humans to manage forests like a crop. Attempts to replicate European conditions elsewhere in the world often failed because foresters lacked suitable native species and found it difficult to grow exotics using European practices.

Second, forestry laws and programs sought to mediate against the worst of environmental destruction wrought by technology and human avarice by safeguarding forest cover as a means of protecting soil, water, and climate. Not every country placed the same emphasis on conservation, but the idea was implicit within almost all forestry management, from afforestation to conservation. These dual goals guided forestry policy globally until the last decades of the twentieth century and the dawn of the twenty-first century when the proliferation of industrial timber plantations and protected areas challenged the justification of the conservation model. We must first go back to the origins of state forestry in the early modern period to understand how these events unfolded.

States, Security, and Science: The Emergence of Scientific Forest Management in Europe

In 1476, the Republic of Venice, one of Europe's most powerful and wealthy states, passed legislation regulating the harvesting of oak and beech in the republic's communal forests.[8] Other European governments soon followed Venice by creating laws to curtail the destruction of communal forests throughout the early modern period (1500–1750). Governments—the Low Countries (1517), Spain (1518), German states (1530s–1540s), England (1543), Tuscany (1559)—passed state forest legislation in response to perceived and real pressures on woodlands.[9] Wood was required for mining, smelting, charcoal making, heating, construction, and other important tasks in the early modern world. Having come out of a century (1350–1450) of greater wood abundance following the depopulation and economic decline caused in the wake of the Black Death, European elites expressed alarm at what they perceived to be an expanding rate of deforestation.[10] Just how acute were these shortages? A pan-European analysis by Paul Warde suggests that some localities in Europe suffered limited acute timber shortages, but that a continent-wide problem did not emerge until the late eighteenth century during the Napoleonic Wars.[11]

We should not downplay anxiety about resource depletion in the early modern period even though there is growing evidence to suggest that the problem of deforestation was not as dire as officials often suggested. Deforestation received serious attention from government officials and the leading intellectuals throughout the early modern period. Elites genuinely worried about the strategic and economic consequences of deforestation because wood was seen as a necessary tool to pursue overseas expansion and maintain security against rival states. Many European states pursued forest territorialization, the demarcation and appropriation of forests, in order to maintain domestic timber security and to increase revenue required to fund more expensive militaries required to keep up with European rivals on the continent and overseas.

The gradual expansion of state power, the extinguishing of lingering medieval rights and privileges of commoners, and the enclosure of privately owned land erupted into social conflicts across Europe.[12] Protests in Europe during the 1700s and early 1800s presaged revolts elsewhere in the world during the nineteenth and twentieth centuries. The transitions that led to these revolts have drawn the attention of historians, who have demonstrated how the imposition of state forestry transformed the way in which Europeans engaged with forests.[13] Yet we must be careful not to see forestry as being the only, or even the primary, cause of the loss of communal land tenure. Forced removals from previously common lands, the intensification of private property, and the commodification of agrarian production were all part of a major shift in the socio-property and economic order of Europe since at least the 1300s that eroded common rights. The origins and causes of this wider process remain shrouded in controversy, but the trend itself was apparent throughout Europe, despite unique local variations and histories.[14] Whether the means (centralization of control and state management of forests) justified the end (forest protection), or whether the "problem" (deforestation) necessitated state intervention, are highly contested questions that cannot be resolved easily.

The origin of modern forest science and management is usually traced back to German-speaking states during the eighteenth and early nineteenth centuries.[15] Scholars in multiple fields have often written (somewhat erroneously) that the concept of the sustainable yield appeared in the 1713 forestry treatise *Sylvicultura Oeconomica* (Economics of Forestry), written by a mining engineer and state official, Hans Carl von Carlowitz.[16] Carlowitz offered recommendations for "artificially" (i.e., using human effort) replanting trees in order to ensure "sustainable" timber production, a translation of the term "*Nachhaltigkeit.*" Although Carlowitz was not the first to discuss the concept of the sustainable harvesting of wood, his ideas proved influential for a succession of German forestry thinkers who developed systematic methods for managing forests in the late eighteenth and early nineteenth centuries. The argument that a sustainable system of forestry could provide a perpetual supply of timber appealed to German state officials who were influenced by the political and economic ideas associated with cameralism. Cameralism was a "state science" devised by political advisors who wanted to modernize governments by compiling economic information—such as statistics about crop productions, the counting of trees in a forest, etc.—to guide policy decisions.[17]

Forestry fit neatly into the cameralist framework because trees could be counted and managed according to the new economic tools of quantification and calculation.[18] German scientific foresters tried to understand the growth rates of trees in forests.[19] Foresters used information on growth rates to devise "working plans" that divided native forests into sections to be cut sequentially when trees reached their economic maturity. Foresters aimed to produce what they described as a "normal forest," that is, a forest with an even distribution of trees of various age groups that could be harvested in perpetuity on a calculable schedule. After harvesting native forests, frequently using a clear-felling method that cut down all standing timber, foresters would encourage "natural" regeneration either by retaining trees to reseed or, more commonly, by creating a plantation by "artificially" replanting with a single commercially valuable species. Plantations became the preferred form of replanting, when possible, because it was a more efficient way to grow commercially productive forests. Plantations also clearly demarcated forests that were being managed "scientifically," and thus, were the domain of the state and professional foresters.

Germans pioneered the profession and science of forestry in part because a number of key innovators—Carlowitz, George Hartig, Heinrich Cotta, to name a few of the most important—wrote down their ideas in textbooks and taught these ideas in the first forestry schools that appeared throughout Germany from the 1780s to the 1790s. Germanic theories of forest management spread throughout Europe as countries—Russia (1803), Austria (1805), Hungary (1807), Poland (1816), France (1824), Sweden (1828), Ukraine (1840), Spain (1848) and Switzerland (1855)—established forestry schools and invited German or German-trained foresters to teach students. As a result of this influence, Germany is often viewed as the fountainhead of European forestry.[20] After influencing forestry regimes in Europe, German foresters and forestry ideas influenced the rise of state forestry outside Europe in the second half of the nineteenth century. James Scott argues that at "the end of the nineteenth century, German forestry science was hegemonic."[21]

German foresters played a significant role in the history of forestry in Europe and elsewhere in the world, but there are some reasons to question the narratives about the uniqueness and hegemony of German forestry. The claim about the centrality of Germany's role relies on the belief in German exceptionalism, a narrative that German foresters themselves advocated throughout the nineteenth century. Interestingly, the basic narrative of the history of forestry—from the eighteenth-century "founding" of the sustainable yield to the global spread of German forest science—has not fundamentally changed since nineteenth-century German foresters began writing the first histories of forestry.[22] German foresters constructed a story of exceptionalism when they demarcated the boundaries of the forestry profession in a bid to justify their expertise compared with nonscientists and other experts. They distanced forestry from closely related scientific disciplines (botany), management techniques (agriculture), and professions (engineering). These narratives are also disconnected from wider European and European colonial history. Historians have perpetuated this self-fulfilling professional prophesy by focusing on the same iconic German foresters and their ideas. The main difference between German foresters and historians is that nineteenth-century foresters imagined that the spread of forestry was good whereas historians tend to see it as bad.[23]

Seeing forestry as a distinctly German science and management practices requires us to artificially separate the science and management of forests from other forms of land management and sciences that developed in Europe and Europe's colonies during the eighteenth century. One of the myths perpetuating this division is the argument that the concept of the sustainable yield arose and developed *specifically* within the discipline of forestry. Yet research indicates that conceptions similar to the sustainable yield principle appeared, independently, throughout western Europe during the seventeenth and eighteenth century as part of wider concerns about managing the organic economy.[24] Of course, forests presented unique problems, especially with regard to the extremely long growth rates of trees (often over a hundred years), which allowed for the concept of sustainable yield to be expressed in terms specific to forests, but the concept of taking no more than could grow back within a set period of time applied to agriculture and husbandry. Sustainability was, so to speak, "in the air" in the early modern era because of pressures associated with a growing population, competition between European states, and limitations (real and perceived) of natural resources.

Plantations are an excellent example of the wider revolution in natural management that occurred in the late early modern era. Sidney Mintz notes that "the history of European societies has in certain ways paralleled that of the plantation."[25] What he means is that the intensification of land and labor in the eighteenth century reflected the growth of a proto-industrial social and economic order that shaped every part of the world influenced by Europe. The modern plantation originated as Europeans sought to rationalize production and "improve" the yield of organic productions—from forests to agricultural crops to minerals. It is this rationalization that James Scott identified in his book, and forestry was just one example that he likely never meant to isolate from other forms of management (as has been done by scholars citing his work). There is even some room to question whether the particular form of rationalization embodied in the plantation was distinctly "western" or a product of the Enlightenment. For instance, Japanese agriculturalists in Tokogawa Japan (late 1600s and 1700s) also created timber plantations to augment declining native timber.[26] This development was somewhat unconnected to European plantations, but it clearly demonstrates that there was nothing particularly exceptional about the use of plantations in central Europe.

Because many historians equate forestry with monoculture plantations, they have rarely examined other forest management practices, such as understory plantings or natural regeneration of mixed-forests, that present a more complex historical story. One exception to this trend is the German environmental historian Joakam Radkau. He notes that "Prussian dogmatism [i.e. rigid plantations and theoretical purity] . . . was counterbalanced by other forestry schools and approaches that come out of the mixed forests of central and southwest Germany."[27] It is not easy to generalize about European forestry because the continent's variety of ecosystems and the different timber needs of regions produced distinct management models suited to particular locales. For instance, in northern France foresters aimed to produce mixed-oak forests rather than monoculture conifer plantations.[28]

Historical claims that imply that timber plantations in central Europe (or elsewhere globally) were an ecological disaster are somewhat overblown. Scholars tend to point to the declining productivity, disease, and death of conifer plantations in the late nineteenth and early twentieth centuries.[29] While the decline of plantations did happen in certain locales, these problems were confined to limited sites and second-generation plantations, rather than being a serious problem that destroyed vast forests across swaths of Europe. Seminatural and plantation forests that were first established during the nineteenth and twentieth century currently constitute 96 percent of Europe's forests, so one can hardly say that central Europe's forests collapsed.[30] In fact, we know now that monocultures are often unhealthy because foresters quickly recognised the broad ecological and biological requirements of forests. The failure of plantations led foresters to allow for the gradual ecological diversification of plantations as a result of purposeful and accidental species introductions. There have been various programs to encourage greater forest diversity throughout

Figure 1.2

Mixed scotch pine, beech, and basswood plantation, Germany
Credit: Photo courtesy of the Forest History Society, Durham, N.C.

Europe in the twentieth century that have succeeded in reintroducing and expanding vulnerable flora and fauna populations. By no means were the forests created in the nineteenth and twentieth century the same as they were in 1500 AD or 1000 BCE, but they have been resilient, enduring, and adaptable to the changing needs of humans and the environment. This is not to say Europe's forests are therefore in an "ideal" state, but the idea that nature has an "essence" is a human concept that should be treated with some caution.

Applying modern conceptions of biological diversity to the practices of foresters in the past is useful from an historical ecology perspective—that is, it helps us reconstruct the past composition of forests—but it fails as a lesson of morality because it assumes historical actors had knowledge of present-day scientific concepts. Using the best available information, foresters throughout central and northern Europe believed that native forests were dominated by a handful of commercially valuable species so they saw little distinction between plantations composed of native trees and native forests. They sought to create forests with native species, rather than importing exotics.[31] Instead of being static, foresters acknowledged that they would experience failure and learn from experience. Heinrich Cotta advised foresters to constantly reevaluate existing practices and theories because detailed knowledge of forests and conditions could only be learned over many generations.[32] After seeing the failure of conifer plantations, German foresters by the beginning of the twentieth century started to pay keen attention to properly matching species to soils, allowing for the growth of nonvaluable species within plantations, and advocating the greater use of mixed forest management systems.[33]

A Europe-wide analysis of the adoption of state forestry programs suggests that national factors—timber scarcity, perception of the environmental problems associated with deforestation, strategic concerns, and land-use—led governments to adopt scientific forestry programs. Switzerland implemented a forestry law in 1876 after examples of flooding in 1868 were attributed to deforestation. As I discuss in the next section, concerns that deforestation caused significant environmental problems first emerged from examples in European colonies (not Germany) before gaining scientific credibility in Europe. Sweden passed legislation in 1903 only when native timber supplies began to dwindle. Instead of "taking over" forests the Swedish state encouraged private owners to afforest former farmlands and heathland through community-based, county forest boards. As a result, for much of the twentieth century, the Swedish state managed only 25 percent of Sweden's forested land, with the rest managed by smaller or midsize private owners (50 percent) and industry (25 percent).[34] Britain was the laggard of Europe because most forests were privately owned. It passed the first significant piece of forestry legislation, the Forestry Act of 1919, only after World War I ended. Between the interwar years, Britain's government funded a large-scale planting program to ensure it had enough timber in case of a future war.

The rise of state forestry and innovations in forestry science helped reverse four hundred years of declining wood production across Europe. In doing so, Europe experienced the world's first "forest transition," a transition from deforestation to reforestation driven by political, economic, and social factors.[35] Countries such as Germany, England, France, Sweden, Finland, and Switzerland all experienced forest transitions as a result of changes implemented in the nineteenth and early twentieth centuries that led to the planting of trees and the protection of existing forests. Changes within Europe's economy—urbanization, industrialization, and the growing use of coal—aided this shift, although the application of conscious state policies to limit access to common forests and to apply scientific management to forest production was one of the primary reasons for this forest transition.[36]

Climate and Colonial Expansion: The Global Conservation Movement

While forests expanded throughout much of Europe in the nineteenth century, many forests elsewhere in the world suffered from an onslaught of deforestation caused by European imperialism and settlement. In 1864, George Perkins Marsh, a former U.S. congressman, described the ravages wrought by colonization: "Comparatively short as is the period through which the colonization of foreign lands by European emigrants extends, great, and, it is to be feared, sometimes irreparable, injury has been already done in the various processes by which man seeks to subjugate the virgin earth."[37] Stories of environmental calamity poured in from around the world. Scientists and scholars became increasingly troubled because they observed that deforestation changed climate for the worse, destroyed soil fertility, and created water scarcity where previously there was abundance. This particular view is described as "desiccation theory" because the theory posits that loss of vegetation cover leads to the progressive drying out of the atmosphere and soil, causing desiccation. Foresters and conservation popularizers, such as Marsh, lobbied governments throughout the world to establish forestry departments to protect forests in order to conserve water, soil, and climate.[38]

In response to these concerns, governments throughout the world established forestry legislation and created forestry services staffed by professional foresters to manage forests using scientific methods between 1850 and the 1920s. Diana Davis writes that "so strong were beliefs in the theory of desiccation that policies were enacted in many European colonies in Africa, Asia, and island territories to plant trees and to conserve remaining woods in order to attract rain, prevent drought, and ameliorate the climate."[39] Independent governments such as the United States and Japan also established modern forestry services and laws. Algeria (1851), India (1864), Java (1865), The Cape Colony (1881), Madagascar (1881), South Australia (1882), New South Wales (1882), Malaya (1888), the United States of America (1891), Hong Kong (1895), Thailand (1896),

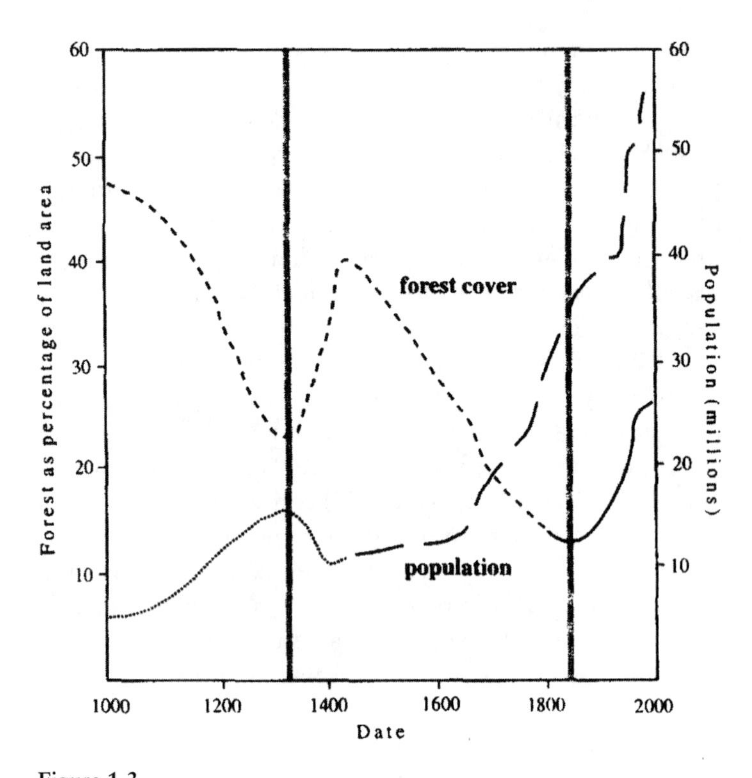

Figure 1.3

Forest transition in France: Population and wood cover

Source: A. S. Mather et al., "The Course and Drivers of the Forest Transition: The Case of France," *Journal of Rural Studies* 15, no. 1 (1999): 65–90.

Japan (1897), South Africa (1910), the Philippines (1904), Canada (1906/1911) Western Australia (1916), China (1916), Morocco (1917), and Brazil (1925) passed forest legislation and established forestry services.[40] The rapid spread of forestry throughout the world reflected the "discovery" that deforestation could change climate for the worse, destroy soil fertility, and create water scarcity.

Humans have speculated on the links between deforestation and climate change at least since the ancient Greeks, although the idea only became the subject of sustained intellectual inquiry in the mid-eighteenth century. The question about the relationship between forests and climate gained renewed importance during the Enlightenment when naturalists sought to understand the relationship between the physical environment and climate. No person did more to shape scientific attitudes about climate during the Enlightenment than Georges-Louis Leclerc, also known as Comte de Buffon, France's most eminent naturalist. Buffon argued that North America had hotter summers and colder winters than Europe because of its large forests. Accordingly, many Enlightenment thinkers believed that the conversion of America's vast forests into farmland was a *good* thing because deforestation moderated temperatures and, thus, aided human settlement. The future American president, Thomas Jefferson, believed that the destruction of forests was improving America's climate for the better. In 1782 he wrote, "A change in our climate, however, is taking place very sensibly. Both heats and colds are become much more moderate within the memory even of the middle-aged."[41] Whether or not the American climate had changed for the better as a result of deforestation remained an ongoing debate well into the nineteenth century.[42]

Naturalists living in the tropics began to paint a dire picture of the consequences of deforestation. European colonization on the islands of Madera, St. Helena, Mauritius, and in the Caribbean raised the troubling possibility that deforestation itself caused negative climatic effects, including diminished rainfall and increased extreme temperatures.[43] The environmental historian Richard Grove notes: "In the tropical zones . . . the amelioration, warming or rainfall reduction which might seem desirable in North America could clearly threaten agricultural" interests.[44] Islands offered ideal laboratories to observe changing climatic and ecological conditions. Officials and naturalists who spent time on the tropical islands became alarmed at the conversion of tropical forest into plantations and denuded hillsides. European colonialists independently observed that deforestation caused declining rainfall and created harsher climates. There was one notable attempt in Mauritius to institute a forest conservancy regime to preserve the climate, but the idea for the most part had little policy influence because colonial officials sought to encourage, rather than hinder, economic development.

Scientific opinions about human-caused climate change shifted in the first half of the nineteenth century. Scientists began to believe that the loss of forests created a cascade of negative consequences, beginning with the erosion of soil, the drying out of water supplies, declining rainfall, before finally ending with drought and desertification. The Prussian aristocrat Alexander von Humboldt offered a coherent physical explanation for climate change and he also documented supposed changes to rainfall regimes and water levels as a result of deforestation in South America.[45] From 1799 to 1804, Humboldt traveled throughout South America making detailed observations about the region that led him to posit new theories on geography, climate, and biology. While traveling near Lake Valencia in modern-day Venezuela, Humboldt observed the rapid destruction of woods to fuel the mines. He also noted the proliferation of plantations higher on the sides of the valley. Humboldt connected declining rainfall with farmers who destroyed native forests to make way for farms. He also argued that deforestation led to diminished streamflow and a lower lake level. In the 1820s, Humboldt encouraged two young scientists Jean-Baptiste Boussingault (1802–87) and Mariano de Rivero (1798–1857) to revisit sites in Latin America that he described in his narratives. When they visited the lake, they found it even lower, a fact that in their minds confirmed

Humboldt's thesis that deforestation caused lower rainfall and increased erosion.[46]

Changes to France's political and physical landscape after the Revolution provided European precedents that agreed with Humboldt's theory. The relaxing of forestry laws during the French Revolution encouraged the destruction of forests by angry peasants who attacked forest guards and disobeyed rules banning access in forests. In the aftermath of this destruction, the French aristocrat Jean-Baptiste Rougier de La Bergerie proposed that deforestation during the Revolution caused water springs to dry out.[47] Later in the century other naturalists, including Boussingault, offered support for "sponge theory," the idea that forests enhance water retention and release water equitably throughout the year rather than concentrating runoff after rain events. Adherents of this theory found evidence in 1830 (and later again in 1868) when horrible floods visibly eroded hillsides and valleys in the Alps. Foresters claimed that the floods resulted from denudation of hillsides by local residents. This claim met with criticism from French engineers, who argued that forests had little hydrological impact on extreme flooding events, but foresters won the argument. Conserving water resources in catchments became one of the key justifications for expanding forest conservation in mountainous regions globally. Along with desiccation theory, sponge theory created a justification for management of catchment and mountainous areas that lasted into the twentieth century.[48]

Anecdotal evidence provided powerful examples that convinced many people of the need to regulate forests for the greater good. The most powerful anecdotal proof about the dire consequences of deforestation came from classical literature. Armed with passages from the Bible and ancient texts, writers from around the world offered a litany of examples of how deforestation in the Middle East, India, and the Mediterranean led to the destruction of storied civilizations and turned fertile land into useless desert. Middle-class Christians in Britain and the United States read popular geographies of the Holy Land that described these changes in detail. "Biblical Botany," an article written for middle-class Americans in the Midwest, told its readers about the drying out of the Holy Land and the entire cradle of civilization: "A . . . mournful change . . . has passed over nearly all the seats of ancient civilization, so that the glowing descriptions of their poets, historians, and geographers no longer apply to them."[49]

What in many instances amounted to a rabble of random passages gained scientific-like status and wide readership because of the tireless work of the aforementioned George Perkins Marsh, author of the classic book *Man and Nature: Or, Physical Geography as Modified by Human Action.* Marsh was a prodigious reader and scholar who compiled a systematic analysis of classical references and scientific findings from around the world. *Man and Nature* put desiccation theory on firm footing by tracing how human action had modified the geography, physical landscape, and climate throughout the world from the classical world until the Victorian era. Marsh wrote his book for "the general intelligence of educated, observing, and thinking men," an appeal to the middle classes that signaled the widening public appeal of conservation.[50]

Marsh's book helped jump-start the conservation movement, a global movement led by advocates who wanted to apply scientific principles to manage natural resources for the greater good of society. This movement had adherents throughout the world, but it gained a particularly strong following in the colonies of Anglo settlement—Australia, New Zealand, South Africa, Canada, and the United States—in response to the series of booms and busts of colonial expansion that left visible scars throughout the world.[51] Surging settler birthrates, the expulsion of indigenous peoples from their traditional lands, and the exploitative use of resources on the frontiers of expansion reshaped human-forest relationships throughout the colonial world. Settlers in Australia cut down so many trees to make farms and sheep stations in the forested southeastern region of the continent that one famous Australian historian wrote that settlers must have "hated trees."[52] Settlers outside the Anglo world experienced similar booms and busts. The coastal Atlantic forests of Brazil were devastated to make way for plantations and to fuel urban expansion.[53]

Conservation advocates sought to curtail the worst excesses of resource extraction caused by the rapacious use of natural resources. They argued that capitalism and the self-interest of individuals required restraining, and only a strong state guided by scientific principles could achieve that goal. Marsh wrote that "abundant experience has shown that no legislation can secure the permanence of the forests in private hands."[54] This view was only reinforced by events during the second half of the nineteenth century. Examples in the American South of landowners who purchased land, cut down all the trees, sold them, and then abandoned the land to move to new forests in the West provided a tangible example of capitalism gone awry. In the United States, the experience of deforestation made a deep imprint on even the wealthiest, most influential Americans. The Progressive conservation movement in the United States was led by elites, such as Theodore Roosevelt and Pinchot, with considerable wealth and power.[55] The conservation impulse was strongest in locations where European expansion had exposed forests to the devastating forces of capitalism, industry, and settlement.

The desire to protect forests was not purely economic or utilitarian in its focus. Protecting forests was the centerpiece of a wider conservation impulse that expanded to include the protection of superlative sites of natural beauty to showcase untouched "wilderness" for the future enjoyment of humans.

Today we recognize that the spaces that people conceived of as being "wild" were inhabited by humans and had been ecologically transformed and even managed.[56] But the belief that wilderness existed, and could be protected, provided a powerful psychological tool in encouraging popular participation in setting aside forests in European settler societies. The establishment of Yellowstone National Park by the United States Congress on March 1, 1872, set off a global movement to declare national parks. Governments in Australia, Canada, South Africa, and New Zealand all declared national parks, including the National Park in New South Wales (1879), Rocky Mountain National Park in Canada (1885), Tongariro National Park in New Zealand (1887), and Kruger National Park in South Africa (1926). Like forestry, there was no single "American" origin of national parks, although like Germany, the precedent for national parks in the United States proved to be a potent justification for establishing park systems.[57]

Outside Anglo settler societies, where governments had appropriated large tracts of land from indigenous people by dispossession or treaty, the concept of the national parks proved more difficult to implement. Parks advocates found it more difficult to apply the principles of national parks in continental Europe, where land had been managed intensively for hundreds of years. Sandra Chaney argues that German advocates "believed that only countries with vast 'uninhabited wilderness' could support such use of land."[58] In 1902, the Polish German forester Heinrich von Salisch wrote that central Europe lacked the "primeval" forests required to construct such a park.[59] Unsurprisingly, then, the first European park, Sarek National Park (1909), was located in Sweden's remote northern region of Lapland. Europe's second national park, Swiss National Park in Switzerland (1914), was established in the rugged Alps. Formal European colonies delayed even longer in creating national parks. India's first park was established in 1936. The Dutch began establishing nature reserves in Indonesia in the 1920s. This delay can be attributed to the lack of middle-class and elite public pressure for parks.

Preservation sentiments flowed from, and merged with, the desire to protect forests for conservation purposes. Environmental historians at one time distinguished conservation and preservation as two distinct movements, although historians now tend to portray the two trends as being extremes on a conservation spectrum. A recent study of the American conservation movement by Ian Tyrell notes the following: "Actors in the debates over conservation did not see a hard and fast choice between conservation for future use, on the one hand, and preservation, on the other . . . supporters insisted that wild animals would flourish in the forest resources . . . Other ardent nature lovers saw national forests as themselves impressively scenic, even when destined for future use . . . Moreover, this sensibility was not American alone so much as international, and it involved transnational influences upon American conservation."[60]

Making a distinction between preservation and conservation denies the strong ties between foresters and park advocates. John Muir, America's most recognized advocate of national parks, believed that forest reserves and national parks were part of the same broad framework of management. He concluded his 1895 keynote speech "The National Parks and Forest Reservations" at the Sierra Club annual meeting by calling on the U.S. government to scientifically manage its vast forests: "Forest management must be put on a rational, permanent scientific basis, as in every other civilized country."[61] Muir, like many other conservation advocates in the late nineteenth century, wanted to protect forests for the wide benefits—economic, conservation, aesthetic, physiological—that they provided. These broader conservation justifications, rather than narrow economic interests, explain why forest management helped propel forward the global conservation movement in the second half of the nineteenth century.

The Conservation Model: European and Global Variations

This section explores the commonalities and divergences shaping forest management globally from the mid-nineteenth to the mid-twentieth centuries, a period that saw the creation of a common set of ideals that governments and foresters used to guide state forest policy. Conservation advocates worldwide believed that states should employ professional foresters to manage forests to ensure the sustainable production of timber as well as the conservation of water, soil, and even climate. Yet the emphasis placed on conservation and timber production varied depending on environmental conditions and political, social, and economic contexts.

It is useful to distinguish broadly between a "European trajectory" in central and northern Europe, and a "non-European trajectory" in countries outside Europe that adopted forestry regimes in the late nineteenth and early twentieth centuries. Foresters in most countries outside Europe faced distinct problems associated with establishing conservancy regimes from the ground up. They had to draft legislation, create new forest boundaries and rules, learn about the ecological composition of forests and the biology of key species, map new terrain, engage with forest residents and other interest groups, and begin to manage forests for production and conservation purposes. The difficulty faced in trying to understand new environments, settle rights, and engage with competing interest groups limited foresters' ability to manipulate forests to the same degree as their European counterparts did until the middle of the twentieth century.

Distinguishing between a European trajectory and a non-European trajectory helps address problems arising from the commonly maintained view that German forestry was replicated everywhere in the world. Scott argues that "the German model of intensive commercial forestry became standard throughout the world."[62] Taken to its extreme, scholars posit that there was almost no difference between German forestry and forestry programs elsewhere in the world. Ravi Rajan has argued that there was "nothing unique or imperial about forestry in the British Empire" because it reflected continental values derived from German precedents.[63] This echoes an earlier argument by Indian scholars Madhav Gadgil and Ramachandra Guha that "[it is] now well established that the imperatives of colonial forestry were essentially commercial" and lacking "broader social or environmental considerations."[64]

Scholars who claim that a strain of European forestry was exported globally have paid particular attention to British India because the Government of India created a vast system of forest conservancy during the mid-nineteenth century. By 1864 the Government of India had developed a framework of forest management for the whole subcontinent. Indian officials successively brought in three Germans—Dietrich Brandis, William Schlich, and Berthold Ribbentrop—to fill the role of inspector general of the forests in India. Indian foresters quickly gained global renown for their scientific expertise, legal acumen, and professional discipline.

What was particularly "German" about the forestry program established in India? Rajan explains that Indian forestry was Germanic because

> the objectives of forest management involved ensuring protection of forests against natural and human destruction; devising a good system to secure the regeneration of the forests, either naturally or artificially; the development of a well-considered and methodically arranged form of working; and the installation of good lines of communication to facilitated protection, the working of the forest, and the expansion of produce . . . Implicit in this mandate was the basic principle of

German forestry—sustainable production and harvesting with the long-term needs of the economy at the fore."[65]

Rajan continues by arguing that, in practice, "like German and French forestry, it thus leveled across the inherent diversity of forests . . . It also led eventually to the transformation of these forests into commercially marketable monocultures."[66] At face value, one sees little difference between Scott's description of Prussian plantations and colonial Indian forestry: foresters in both regions turned diverse forests into monoculture plantations to fulfill a Prussian scientific ideal.

Undoubtedly, German foresters played a role in shaping forestry theory and practice throughout much of the world. But this alone does not mean that forestry was Germanic in form. There are two significant problems with arguments that forest management around the world followed German precedents, and both can be illuminated by the previous example. First, scholars tend to rely on a description of "German" forestry that lacks precision or detail. Is installing good lines of communication really German? The Indian Forest Service was different from other government departments in India in that it had distinct hierarchy and clear mechanisms for communicating across the vast subcontinent. How can protecting forests be described as German? Naturalists and government officials had argued for the protection of forests in India for strategic and climatic purposes for a half century *before* German foresters ever visited. As I discussed earlier, many of the supposed qualities that historians describe as being "German"—from plantations to the desire to manage forests—reflected common attitudes then prevalent throughout Western and some non-Western cultures of the time.

Second, the argument about the Germanification of the world's forests assumes that foresters tried, and then succeeded, in turning previously diverse forests into monoculture plantations. This position often assumes that the imposition of forestry destroyed precolonial ecological equilibriums characterized by higher levels of biodiversity and a more sustainable relationship between indigenous people and nature. There has been pushback against the idea that idealized precolonial ecological equilibrium conditions existed throughout many regions of the world, but little attention has been directed to the claim that foresters around the world turned native forests into monoculture plantations.[68] Many forest ecosystems had already been modified, sometimes significantly, by human action before the onset of state forestry.[69] In reality, foresters establishing new regimes in other parts of the world lacked the knowledge, capital, labor, and technology to turn native forests into plantations. This is especially true in tropical climates. A commentator in 1911 noted that forestry work in India "thus far has been chiefly one of organization, survey, and protection" without significant advances in utilizing or regenerating forests.[70] Edward Stebbing, a former inspector general of forestry, wrote in his 1923 history of Indian forestry that "to imagine . . . that the devastated forests [of India] could be replaced by plantations in so large a country as India . . . was a fallacy."[71] The wider history of plantations is discussed in more detail in the next chapter, although it is worth noting that prior to the mid-twentieth-century plantations outside Europe were not widespread or particularly successful.[72]

The point to be inferred from this discussion is that historians must pay more attention to on-the-ground constraints that limited the applicability of theoretical ideas, be they German or Indian, when writing about forest management systems across the world. Based on his study of Tanzanian forestry, Thaddeus Sunseri warns, "It would be incorrect to reify the power of the state, capital, and colonial science as forces capable of transforming forest relations according to the European template."[73] His view is confirmed by other research that indicates that state policy, management practices, professional cultures, and scientific theories differed greatly from country to country, and even within distinct regions of a country. This research also suggests that foresters, in Europe and elsewhere, often rebelled against the idea that foresters should aim to turn native forests into monoculture plantations.[74] Judging from theory and practice, the history of forestry cannot be seen as one big attempt to turn the world's forests into plantations.

Any characterization of global forest management must first take into account the political, social, economic, and environmental constraints limiting the power of professional foresters. Many historians give foresters an almost hegemonic-like power. This arises because scholars have devoted significant attention to the social conflicts that occurred when foresters imposed new state legislation on forest dwellers. These trends are important to understand, but by focusing selectively on these instances, historians have given foresters more agency than they actually had. Though foresters succeeded generally in convincing governments to create state forestry programs, foresters themselves had little power. Foresters usually wanted more forests and greater control over them, but they rarely received these requests in the face of competing claims by industry, local residents, and other government agencies. Although forestry may have been a romantic career ideal for many youth dreaming of the open air and green forests, in reality foresters were usually subservient to officials working in higher levels of the civil service. In the colonial world, foresters fell well below district officers, judges, and other senior officials. Foresters had little political influence compared with major industries or lobby groups.

Advocates of forestry skillfully utilized foreign precedents as rhetorical examples to convince elites and the public to adopt a program of action, but in reality the creation of state forestry systems reflected local circumstances. Tyrell points out that "it was not so much the international influence on technical training that was important as the inspiration provided by German forestry."[75] Successful forestry advocates like Pinchot in the

United States, Dietrich Brandis in India, and David Ernest Hutchins in South Africa drew on European precedent when required, but they also argued frequently that German practices were, on the whole, not applicable to local conditions.[76] Pinchot, America's first chief forester, noted that "admirable as German Forestry certainly was, there was about it too much artificial finish, too much striving for detailed perfection."[77] Brandis actually had no training in forestry, something that helped him to respond flexibly to the varying social and ecological conditions in India. On the contrary, foresters who rigidly attempted to import European methods like Comte de Vasselot in the Cape Colony, Charles Edward Lane Poole in South Africa and Western Australia, and Fernow in the United States and Canada experienced erratic, often unfulfilled careers because they quickly made enemies or found themselves unable to work effectively within the context of local politics and social conditions.[78]

Foresters served a key role as arbiters who tried to find common ground between public and private interests. To that effect, they worked closely with private industry but did not want the forest industry to determine how state forests were managed. Private industry ranged from large corporations that produced and processed wood to small teams of woodcutters. Though small teams continued to harvest timber in state forest systems to supply local mills and small-scale wood industries, large vertically integrated companies increasingly dominated the growing and manufacturing of timber products across the twentieth century. Larger firms amalgamated smaller timber operations and had the capital to purchase new technology to harvest and process a variety of timbers. The transition from smaller firms to larger mirrors general business trends and also happened in the timber industry because of the high cost to develop and utilize new technologies required to turn timber into industrial products.[79] Today, the timber industry is dominated by a few dozen multinational companies based in the United States, Finland, Sweden, Germany, Brazil, China, Canada, and Indonesia.[80] These companies increasingly source timber from privately owned plantations, although many rely on timber harvested from government forests.

A number of immediate problems confronted the first generation of professional foresters employed by states. They faced the reality that many of the most valuable forests had already been destroyed. The Australian state of Victoria established a Forestry Commission in 1918, well after many of the state's most valuable and largest forests had been destroyed by settlers and the mining industry who sought, as Tom Griffiths notes, to turn "timber to tinder" to make way for farms.[81] New Zealand, Australia, South Africa, Brazil, and Canada all experienced high rates of deforestation before the imposition of a state forest conservancy. The forests that were still intact were often located in higher mountains or far remote areas that had presented accessibility challenges for would-be settlers, miners, and loggers. Mountains protected forests in the United States, Canada, parts of Australia (like Victoria), and New Zealand. Dense rain forest and jungle saved many of the larger forests in tropical Latin America, central Africa, and South and Southeast Asia from early destruction.

Contrary to depictions that the first generation of professional foresters tried to replicate German plantations, as a rule foresters focused their earliest efforts on conserving and expanding existing native forests. This meant in the first instance that foresters needed to demarcate boundaries of state forests and to determine the future privileges and rights of local forest residents and the general public. Making boundaries in forests created stronger distinctions between state and private property, thus extinguishing or limiting many traditional rights that people had previously enjoyed. When possible, foresters compromised by giving certain privileges (they tried not to define them as rights so as not to perpetuate them), but many locals believed these privileges were too narrow or restrictive.

The concept that states and foresters employed to handle these conflicts can broadly be described as "multiple-use," although in the American context this has a specific legal meaning and history.[82] Multiple-use meant that the public retained some rights or privileges within state forests, though the extent to which the public could exercise these rights varied. In European settler societies, white landholders and private industries often received greater rights because they could influence the political process and the drafting of legislation. Grazers and miners generally received some type of access to minerals and grazing on government-owned forestland. Multiple-use policies also encouraged rising tourism in national forest systems throughout the twentieth century in European settler societies.

Multiple-use took on an entirely different character in formal colonies and from the perspective of indigenous people in settler societies. Indigenous people in both systems had less ability to express themselves through direct political action, something that gave the state greater power in determining forestry regulations. Colonial governments tried to balance the timber needs of indigenous communities with demands on timber by other, more powerful lobbying groups, such as the state, military, and European businesses and industry. Colonial foresters struggled to work with other colonial officials and indigenous residents to construct regulatory regimes that accommodated indigenous users' existing access to forests.[83] Yet foresters often found that their efforts to punish people for breaking forestry laws were in vain. European magistrates and district officers frequently sided with indigenous people against forestry officials to keep the peace because forestry was highly unpopular among people living near forests. Despite receiving criticism from some European officials, colonial forestry regimes assumed that indigenous groups were selfish wood users without any sense of conservation.[84] Foresters pointed to "supposed" examples of widespread deforestation in the Arab Middle East, northwest India, and southern Africa to create an archetype of the destructiveness of non-Europeans. To keep indigenous people from "damaging" forests, foresters often forced the removal of people from demarcated forests. Unsurprisingly, foresters and local forest residents often disliked and distrusted each other. There are many reports of local residents who killed or attacked forest guards to protest forestry laws and scare away potential forestry recruits. In many countries, such as those in South and Southeast Asia, there has been a lingering distrust of forestry officials since the imposition of forestry legislation in the last half of the nineteenth century.

Along with controlling access, another pressing problem was trying to exclude destructive fires from burning down forests. Fernow told his students that "the greatest danger to forest . . . is fire, and the protection against this most unnecessary evil, resulting mainly from man's carelessness, absorbs a large part of the energy of the forester."[85] Historians and foresters alike have suggested that the phobia of forest fire originated in Europe, especially France.[86] Even though theories of fire were often influenced by European science, people outside Europe worried about forest fires well before the implementation of scientific forestry management. It was apparent to anyone living near a forest that out-of-control fires burned valuable timber and could even wipe out entire villages and cities. Massive forest fires had been a recurring problem for people living near forests, especially in fire-prone regions in the United States, Australia, and Canada. The most devastating fire in American history, the Peshtigo Fire in 1871, killed sixteen hundred people in Wisconsin. Fire suppression remained a standard government policy throughout most of the world in much of the twentieth century. The success of fire suppression policies in countries such as the United States had the ironic long-term effect of increasing the risk of devastating fires, a link discussed in chapter 3.

With forests safe from fire and people, foresters aimed to conserve forests to produce sustainable timber outputs without destroying the ability of forests to protect water and soil. Foresters were often cautious about opening up native forests to harvesting. They waited decades, sometimes even a century, before they decided to harvest timber in specific forests. In the United States, the forests grew at a faster rate than they were harvested during the first half of the twentieth century because the USFS waited until private landowners had depleted their forests before releasing large quantities of timber from federal forests.[87] This does not mean that foresters wanted to protect old-growth trees from being harvested—in fact, they wanted to cut old-growth trees because they had reached the end of their growth cycle—but these policies did protect many of the world's remaining temperate old-growth forests until societies and scientists became aware of their unique importance.

Foresters tried to regulate harvesting, but the allowable harvest rate of timber in government-controlled forests was usually determined through complex political processes that did not always involve foresters. In the initial phases of forestry, leading foresters often protested government decisions loudly when they felt that the harvest rate threatened the viability of forests. In one famous example, the conservator of Western Australia resigned in disgust in 1922 when the premier of the state set the harvest rate high to help out the British company Millars.[88] Thailand's first conservator, Herbert Slade, was removed from his position in the late 1890s because he criticized the government strongly for logging teak at an unsustainable rate.[89] State forestry services gradually allowed for increased harvesting throughout the twentieth century as they developed better scientific understanding of forests and as governments and society demanded more timber from national forest systems. The criticisms that arose in the last quarter of the twentieth century reflected this increase in timber harvesting.

After foresters protected and mapped forests, they drafted working plans for determining the harvest rate and regeneration of cutover forests. These working plans recommended the methods of harvesting (clear-felling, selective-felling, etc.), the type of regeneration (natural, artificial, etc.), and the time period for harvesting the forests. Prior to the end of World War II, the quality and quantity of these plans varied greatly. It took time for many forestry services to create comprehensive management plans because of a lack of money, staffing, and regulation. Most countries lagged well behind the Germanic ideal of efficiency and control. Some forestry services were slow to develop management plans.[90] The USFS under Pinchot's leadership decentralized decision making, a trend that cut against Germanic emphasis on hierarchy and centralization.[91] Forestry services tended to be somewhat cautious in the initial stages of planning. Pinchot called for "conservative lumbering" in U.S. federal forests. Only after World War II did working plans aim to achieve the goal of "maximum sustainable yield," a concept that is discussed further in chapter 3.

Foresters initially had a built-in distrust of the timber industry based on the devastation of forests globally in the nineteenth century. In 1897, Dietrich Brandis noted that the experiences from this period "taught a lesson, which public men in India have gradually learnt. It is not safe in India at present to entrust the management of public forests to private enterprise."[92] Similar attitudes pervaded forestry services in other parts of the world. There was a warmer relationship between foresters and industry in countries where forestry efforts focused on creating new supplies through afforestation. These countries included Finland, Sweden, Australia, New

Zealand, and South Africa. The balance of power between industry, private landholders, and state foresters differed according to regions and countries. State foresters often wanted to regulate private forests, but this was rarely possible. For example, some leading forestry advocates in the United States called for the regulation of control of privately owned forests, a view that came to prominence after World War II ended, but this push ultimately failed in the early 1950s when Republican president Dwight D. Eisenhower sided with private owners and ruled out this option. In some regions and countries private landowners played a leading role in the production of timber.

Prior to World War II, forestry services lacked the labor, scientific knowledge, and technology to transform diverse forests into monoculture plantations. Financial limitations hindered the ability of foresters to reshape ecosystems on a large scale. Forestry departments, despite their management of large geographic areas, did not receive large amounts of money from the state. Foresters brought in contractors to harvest timber and relied on cheap labor or penal labor to manually plant saplings. General labor shortages made it difficult to replicate densely planted central German spruce plantations. South African plantations, discussed in the next chapter, succeeded in part because of the country's supply of cheap labor. A prominent forester noted in 1935, "South African plantations . . . were only profitable because of cheap labor, indeed, these plantations would not exist without it."[93]

Foresters used a variety of practices to conserve and exploit forests. It is difficult to make generalizations about global practices because they varied widely. But one generalization is easy to make: that foresters mimicked a simplistic "Germanic model" by establishing monoculture plantations cannot be further from the truth. North American foresters working in large mixed native forests frequently relied on some form of the "shelter wood" system, which involved leaving standing mature trees to reseed cutover lands. Shelter wood systems usually allowed for a wider variety of species to regenerate after harvesting. Even when foresters tried to get rid of less desirable species, such as the hemlock in Douglas fir forests, they found it challenging to stop forests from diversifying naturally. In tropical regions foresters sometimes encouraged indigenous farmers to plant seeds of valuable species, especially teak, as part of slash-and-burn agriculture. But as a general rule, most foresters disliked slash-and-burn agriculture, also known as shifting cultivation, because it destroyed forests.[94]

Forestry policies were not merely about producing timber. National forest policies encouraged the protection of key flora and fauna. Thomas Webber, a British forester in India, noted in 1902: "In establishing a Forest Department and protecting the timber from destruction, Government has also extended its protecting arm over the game, so that it shall not be exterminated in a ruthless and wasteful manner."[95] Conservationists sought in particular to protect fauna that could be hunted, such as deer or even tigers. But this is not to say that foresters and other conservationists cared little for other aspects of nature: they supported the creation of preserved parks and wilderness areas within forest systems, something discussed more in chapter 3. The protective functions of state forest systems expanded throughout the twentieth century in response to changing scientific and public attitudes about ecology and the protection of biodiversity. Some of the kernels of current biodiversity management laws and policies, which seek to preserve ecosystems and rare species, are to be found during this early period of forestry.

Forest management policies and practices changed significantly after World War II in response to the growing demand for timber in developed and developing countries. There was an increased need for public forests to produce timber for the

Figure 1.4

Selective-felling hardwood forest in the American South

Credit: U.S. Forest Service photo courtesy of the Forest History Society, Durham, N.C.

market to replace declining production from private lands and from easily accessible forests. Most of the wood consumed outside Europe during the first half of the twentieth century came from native forests. Michael Williams notes that 99 million hectares of "virgin" forest disappeared from temperate regions from 1900 to 1950, much of it caused by logging.[96] Foresters shifted from conservation to production, a shift that would cause significant social conflict beginning in the 1960s. The ramping up of production undermined the claim by foresters that they were conservators of nature. Forest harvesting techniques, especially clear-felling, became more intensive and industrial.

Conclusion

The takeaway point from a survey of forest management techniques is that foresters outside Europe were limited in their ability to transform forest ecosystems into monoculture plantations. In fact, plantations constituted a miniscule percentage of the world's forests until the last quarter of the twentieth century, when the total number and size of plantations increased dramatically, as discussed in chapter 2. Foresters outside Europe were relatively conservative in seeking to harvest in diverse native forests because of the challenges they faced in regenerating commercially viable forests that also produced conservation benefits. Forest management from the late nineteenth century until the beginning of World War II was characterized by a cautious optimism that scientific experts could understand and regulate natural systems for the benefit of humans and many of the key species found in forests. We should not judge foresters for lacking the same sentiments we hold true today because this leads us to overlook the historical evolution in forest management practice and theory that has made a deep imprint on popular and scientific conceptions relating to biological and ecological conservation. Modern conceptions of biodiversity and the importance we now attach to maintaining diversity in forests developed as a result of the processes discussed in the next two chapters, the emergence of intensive plantations and the growth of protected areas.

2

Plantations: From Security to Profitability

Foresters have long dreamed of creating plantations of fast-growing exotic species in tropical countries to provide wood for industry and fuel. *Eucalyptus* proved to be particularly desirable trees because their fast growth and hard timber could be used for fuel, fire, mining and other industrial purposes. Efforts to grow eucalypts in tropical climates dated to the middle of the nineteenth century when foresters planted them with little success. A century later foresters had still not succeeded in their goal, despite considerable efforts. In 1952, the Thai forester Sukhum Thirawat noted, "To tropical foresters Eucalyptus is something of an enigma; a genus so versatile and yet despite considerable attempts at introductions over long years, not one species can be cited as a success anywhere."[1] Finally, foresters discovered species and methods for establishing eucalypt plantations in tropical climates in the 1960s and 1970s. This paved the way for the rapid expansion of exotic plantations, especially eucalypts, in tropical countries.

Ironically, by solving the riddle of eucalypts, foresters created new social and environmental problems. Encouraged by international aid organizations, governments in developing countries began to afforest with eucalyptus to supply pulp for industrial mills and to provide fuel wood for local residents. Although many people welcomed eucalypts because they provided local fuel, the creation of large timber plantations for industrial purposes inspired protests. When the Indian government created eucalyptus plantations in the Karnatka Province, local residents went into the fields and destroyed the young eucalyptus trees, which they claimed decreased local employment options and hurt the livelihood of poor people who relied on agricultural produce.[2] In 1992, a mob of four thousand peasants from northeastern Thailand marched angrily toward Bangkok to protest the Kho Jo Ko resettlement plan, a program implemented by the Thai military junta to remove peasants without legitimate title from government-owned forestland, cut down the forests, and create plantations of exotic species.[3] When the military government fell, angry peasants marched on Bangkok to protest the scheme. In response, the new government negotiated to stop the forcible removal of people and to regulate the harvesting of native forests and the creation of plantations.

Tropical forest plantations now constitute an increasing percentage of the world's forest cover. A review published by the United Nations in 2010 estimated that trees planted by humans covered 264 million hectares of the world's land surface, or approximately 7 percent of the world's total forest cover.[4] Half of those planted forests are located in Asia. Intensive plantations account for only 3 percent of total forest cover but produce approximately 30 percent of the world's industrial wood.[5] That figure is predicted to grow to 50 percent by 2020.[6] The proliferation of timber plantations, especially those in the tropics, reflects global changes to the science, management and utilization of wood since the mid-nineteenth century.

Plantations: Universal Impulse, Local Constraints

Few trees are more recognizable than species from the genus *Eucalyptus*. Eucalypts can be found growing in most tropical or subtropical countries throughout the world. They are frequently visible parts of the landscape in cities and rural areas. Eucalyptus plantations are important contributors to the economies of Brazil, India, China, South Africa, Portugal, Australia, and Thailand, to name just a few countries. Eucalypts have been planted throughout the nineteenth and twentieth centuries for numerous reasons: to provide fuel, timber, shade, and windbreak, and even to cure malaria and tropical diseases. The spread of eucalypts is one example of how forestry knowledge and practice developed in diverse locales around the world and spread through transnational networks to other parts of the world.

The creation of timber plantations can be understood as part of an enduring human desire to manipulate the environment for survival and benefit. While scholars have focused particularly on Germany as the origin of timber plantations, timber plantations emerged independently in East Asia at the same time as a response to perceived and real timber shortages. The development of plantations in central Europe and East Asia shared many similarities: there were perceived resource shortages, and there was an availability of native species that could be grown in plantations. Most of these planting efforts happened in areas that were previously forested, a process known as *reforestation*. Reforestation does not necessarily mean that the ecology of a forest is the same when it regenerates, but it does imply some ecological, structural, or landscape continuity.

A new type of plantation emerged in the mid-nineteenth century in response to European colonial expansion. Colonial governments began pursuing *afforestation*, the planting of trees in treeless areas or regions deficient in certain species and landscape types. The most notable afforestation programs occurred in France, Algeria, India, the United States, and South Africa. Until the early twentieth century many foresters and botanists believed that, without human interference or natural disaster, most of the world would naturally be covered in forest. They saw the existence of vast grasslands and deserts as having been caused by human-induced deforestation. By planting trees, they hoped to increase rainfall and make entire regions more fertile and suitable for human settlement and agriculture. Foresters found it difficult to find appropriate species to plant in these conditions, because there usually were not suitable native species that fulfilled the desired requirements in terms of timber structure, growth rate, and function (e.g., shade, windbreak, etc.). In the past, afforestation was viewed as an "improvement" of nature, but within the past fifty years these plantations have come under increasing criticism.

Many of the most contested plantations are those that involve *exotic reforestation* where a native forest is harvested and replanted with an exotic species. Exotic reforestation occurred primarily in forests where the existing forest was conceived to have a lower economic value than a plantation composed of exotics. Diverse tropical forests, conifer forests in Argentina and Chile and hardwood forests in Australia and South Africa were sometimes cut down to make way for exotic species.

By the 1960s and 1970s, international efforts to find suitable tree species that grew in plantations in the tropics began to succeed. The size and number of timber plantations in tropical countries started to grow from the 1960s on, increasing by approximately half a million hectares per year until the 1980s. Beginning in the late 1980s and early 1990s, the pace of growth exploded. Since the mid-1990s, timber plantations have been growing torridly at approximately 4–5 million hectares per year, with the majority of that growth located in the tropics and subtropics.[7] To understand this proliferation, we must start with the ancient origins of the first efforts by humans to plant trees.

Early Modern Timber Plantations: Regenerating Native Forests

The desire to plant useful species has been an enduring theme throughout human history. Humans have cultivated trees since the dawn of the Neolithic Revolution some ten thousand to eleven thousand years ago. Some scientists think that the domestication of the fig tree even predated the domestication of grains.[8] The desire to plant trees for their wood is a relatively recent phenomenon that began in various parts of the world in the early modern period. Timber scarcity and a growing commercial market for wood products in western Europe and East Asia created conditions favorable to the establishment of plantations. Population pressure in the hill tracts of southern China led enterprising residents in the sixteenth and seventeenth centuries to clear land and establish fast-growing tree plantations composed of indigenous bamboos and the native *Cunninghamia lanceolata.*[9] Independent of China, Japanese agriculturalists produced farm manuals focusing on tree-growing in the 1700s. These manuals influenced the development of Japanese timber plantations in the early nineteenth century.[10] Conrad Totman notes that "such practices [regenerative forestry] arose independently in Japan at least as early as in Germany."[11]

Timber plantations also emerged in central Europe during the early modern period, as discussed in chapter 1. Central European states planted trees for similar reasons: they were concerned about timber shortages and they could sell wood to growing markets that required timber for multiple uses. Like agriculturalists in Japan and entrepreneurs in China, German foresters planted species that were native to the region. Northwest and central Europe's postglacial ecology, fertile soils, and relatively consistent precipitation provided the ideal environmental conditions in which to develop timber plantations.

Europe is home to a relatively small number of tree genera and species compared with many other regions of the world, but the continent is especially rich in conifers such as pine (*Pinus*), fir (*Abies*), larch (*Larix*), and spruce (*Picea*) species. Conifers colonized the land of northern Europe after the retreat of ice sheets during the ending of the last Ice Age. Many conifers are ecological pioneers that thrive in forest openings, such as those that occur after fires, and grow quickly in open sun, crowding out competitors. Many conifers tolerate a range of climatic and soil conditions. They are strong and easily worked, making their timber highly desirable for construction. When Europeans decided to make timber plantations from conifers, they had a range of native species that could be grown fast and easily in monoculture plantations.

Early modern plantations in central Europe, southern China, and Japan were part of a process of *reforestation* involving the replanting of native species that were either indigenous to the site or native to the wider region. Within central Europe tree planting significantly changed the ecological structure of forests. The planting of plantations of conifers reversed central Europe's preexisting ecological balance: conifers became more prominent than broad-leafed hardwood trees, such as oak and beech. Conifers planted by foresters continue to cover the majority of areas where broad-leafed forests once predominated.[12]

The factors that shaped the origins of early modern timber plantations—resource scarcity (real or perceived), state or private efforts to establish plantations, and the availability of native species that tolerated growing in plantations—repeated themselves in other countries later in the twentieth century. This can be explained in part by two factors, the first relating to the human desire to manipulate nature for human needs, and the second relating to biogeographic factors beyond human control. As a general rule, the most productive locales for plantations have been those that also housed large native forests. The world's largest, most productive forests are found in the vast conifer forests that lie between the 40th and 70th parallels in the Northern Hemisphere. As a result of this geographic fortune, foresters throughout temperate regions—including much of the United States and Canada—had fewer problems using native species to regenerate cutover forests. Humans living in areas without abundant natural forests, and those in regions deficient in softwoods and other useful species, found it far more difficult to establish plantations.

The Challenge of Afforestation

Many regions of the world lack forest cover and are deficient in trees that have characteristics most desired by modern humans. Scientists today recognize that forestless areas, such as deserts and grasslands, have lacked forests for hundreds of thousands, even millions of years, for good reasons. Extreme climates, aridity, fire, and the evolutionary adaptation of non-tree species made it difficult for trees to flourish in deserts, heathland, grasslands. All of these environmental factors conspired to make it challenging to establish plantations using native or exotic species without considerable experimentation.

European ideas of forestry did play a key role in shaping attitudes toward tree planting in southern Europe and the European colonies in the second half of the nineteenth century. Foresters argued that regions that lacked large forests, such as the Mediterranean, had suffered from deforestation that denuded forests. They imagined that forests should naturally cover the earth. Scholars describe this as the "ruined landscape theory."[13] This theory was supported by the eighteenth-century idea of improvement, which suggested that humans should transform "wasteland" without agricultural or forestry value into usable forms. It gained renewed interest as a result of desiccation theory, discussed in chapter 1, which argued that forests stopped progressive drying out of landscapes by encouraging rain and maintaining more equitable streamflow.

The most successful nineteenth-century afforestation project was the massive pine plantation that French foresters created in the Landes region of Gascony in southwest France near Bordeaux. Originally a sand dune and swamp-covered landscape, French foresters planted thousands of hectares in the mid-1800s with the maritime pine (*Pinus pinaster*). International forestry boosters drew on the example of the Landes plantations to show that large-scale afforestation efforts could succeed.[14] The precedent of Landes encouraged foresters in other parts of the world to plant trees to provide timber and to improve climate and enhance water conservation.

Foresters and settlers in arid European colonies hoped to replicate the success of Landes. The biggest and most ambitious nineteenth-century tree-planting projects occurred in southern Africa, Algeria, and parts of Australia, regions devoid of dense forests and dominated by arid, hot climates. Millions of trees were planted in these colonies during the last half of the nineteenth century. Enterprising French colonial foresters in North Africa dreamed of "remaking the granaries of Rome" by turning arid deserts into lush, green forests that could be settled and farmed by French colonists.[15] White settlers on the Cape of Good Hope wanted to plant trees to push back farming into the arid Karoo desert. South Australians' first conservator of forests John Ednie Brown told settlers that they could change the interior climate of South Australia by planting trees.[16]

Colonists and foresters alike saw species from the genus *Eucalyptus* as one of the most viable and desirable species to grow in arid conditions. Grandiose claims swirled among international botanical circles about the amazing properties of eucalypts, bolstered by the bombastic writing of the Melbourne

Figure 2.1

Burned and cutover pinelands in Michigan

Credit: U.S. Forest Service, Eastern Region photograph courtesy of the Forest History Society, Durham, N.C.

botanist Ferdinand von Mueller, who championed eucalypts as a wonder tree. He claimed that some species, especially the vaunted Victorian blue gum (*E. globulus*), grew to enormous heights in short periods of time. Many swore that eucalypts helped cure malaria and other tropical diseases because they supposedly excreted antimalarial oil and they had desiccating properties that could drain swamps. Others claimed that Eucalyptus timber was even stronger than teak and oak.

CHANGE OF FORTUNE.—To a scene of his own country combined with an expression of sentiment the Brazilian artist, Weingartner, gives the title of "Change of Fortunè." It represents the newness of a plantation just cleared in the forest of the artist's own section of Brazil, the State of Rio Grande do Sul. The smoke is still rising from the burning brush. Great logs and stumps cumber the ground. Almost surrounding the cleared tract rises the dense forest. In the distance is the cabin. The figures of the foreground are the man and the woman whom change of fortune has prompted to the work of establishing a new home in the wilderness. The appearance of the man on the barrow testifies to an unaccustomed occupation. The attitude is one suggestive of weariness and despondency. The wife has been aiding to the extent of her physical ability, even handling the hoe between the rows of the growing plants. She, too, has stopped. Resting one hand upon the top of the hoe handle, she looks upon the blisters in the palm of the other hand. The woman appeals to sympathetic interest even more than the man does. The painting tells a story of human vicissitudes full of pathos but in colors bright and in a scene which promises better days for the future. The vividness of the Brazilian verdure and the clearness of the atmosphere are not, but the spirit of the scene is, reproduced by the camera.

Figure 2.2

Forest clearing and land settlement in Brazil for exhibition at the World's Fair in St. Louis, 1904

Credit: Photo courtesy of the Forest History Society, Durham, N.C.

Unlike Landes, which still exists, these dreams failed to turn into reality. Such heady dreams lasted for only few decades before fading like a mirage in the desert. Instead of turning deserts into granaries and forests, most seeds never germinated and what trees did grow shriveled, withered, or died before reaching maturity. Out of the millions of seeds that were sown or planted as saplings, foresters could find only a few cool-climate regions, such as the southern highlands in India, where a few species could be grown in plantations. What trees did grow could never live up to their hype. Farmers who listened to Brown and settled in the interior lost everything when drought hit in the mid-1880s; Brown resigned his post and took a job in the neighboring colony of New South Wales. The once ebullient global mood turned to frustration and bitter disappointment.

Without knowing it, foresters plunged headlong into problems. Most of the eucalyptus species planted in tropical and subtropical climates came from the coolest parts of Australia: Tasmania and southern Victoria.[17] Foresters unknowingly planted species far outside their indigenous climatic range and did not have adequate labor to create and tend plantations. The classification of eucalyptus species created an extra problem. Even the leading experts in the world debated their classification, and botanists in Australia had very little knowledge of the geographical range, biological habits, and local ecological conditions in which species grew.

The failure of afforestation projects throughout the world forced foresters to formulate more rigorous experiments for selecting exotic trees and overseeing plantations. Foresters recognized that a number of problems mired attempts to establish plantations before the 1900s. Many exotic timber plantations failed because foresters selected tree species from vastly different climates or received improperly classified seeds. Poor collection methods, especially in Australia, meant that a high percentage of seeds from Australia that foresters planted were incorrectly classified. Matching species to climate was another critical, yet largely overlooked, area of silviculture outside Europe before the 1880s and 1890s. David Hutchins, one of the world's pioneering foresters of the last decades of the century, popularized a simple idea that became a creed for many twentieth-century foresters: "fit the tree to the climate."[18]

The public popularity of tree planting soared during the late nineteenth and early twentieth centuries in settler European countries despite difficulties in establishing timber plantations. Public tree planting was especially popular in the United States, Canada, New Zealand, Australia, and South Africa, where deforestation was a political issue and nationalism was on the rise. Activists promoted tree planting to improve climate, regulate hygiene, and provide a virtuous activity that built a sense of nature and nation. Local and national governments around the world supported tree planting through loans, competitions, and planned tree-planting days, the most famous being Arbor Day.[19] J. Sterling Morton founded Arbor Day in the state of Nebraska to revitalize the state's treeless plains landscape and increase the prosperity and virtue of its residents.

While many European settlers agreed that planting trees was a good thing, people differed on whether to plant natives, which symbolized new nations, or exotics, which had deeper cultural connections and provided a wider aesthetic and ecological palette to choose from. This question proved to be a complex issue for many. In Australia, many colonial nationalists during the 1880s and 1890s promoted the planting of native eucalyptus and wattle to create a sense of "Australian" national identity that was rooted in the Australian landscape, but other settlers disliked Australian native trees because they provided less shade and were not colored as brightly green as many European and North American exotic trees.[20] In the end, most cities and rural areas were planted with a mix of exotics and natives.

State Plantations

The size and economic importance of timber plantations increased in temperate and subtropical climate regions of the world during the 1900s to the 1930s. New Zealand, Australia, South Africa, Brazil, and the United Kingdom developed extensive timber plantations. These countries had negligible plantations prior to the turn of the century. But by the late 1930s, timber plantations covered approximately 300,000 ha in South Africa, 280,000 ha in New Zealand, 200,000 ha in the United Kingdom, and several hundred thousand ha in Brazil.[21] With a few exceptions, such as the growing wattle and eucalyptus industry in South Africa[22] and eucalyptus plantings in Brazil, state forestry departments created and managed the majority of the world's plantations.

These plantations were created in response to environmental and economic pressures, such as deforestation and the effects of World War I, and reflected the pessimistic outlook of foresters at the time. At the dawn of the twentieth century, the environmental damage from settler colonialism was obvious: devastated forests in the Northeastern and Southeastern United States, ghostly white dead eucalypt forests in southeast Australia, burned rain forests in southeastern Brazil, and the recent creation of grasslands of New Zealand where ancient forests once stood were just a few of the most obvious reminders of the rapid deforestation that transformed entire regions of the world. This real destruction reinforced the dire warnings of forestry advocates. One of the greatest timber pessimists was the first head of the USFS, Gifford Pinchot, who warned Americans in 1910 that it was "certain that the United States had already crossed the verge of a timber famine so severe that its blighting effects will be felt in every household in the land."[23] Pinchot's dire predictions became a brief reality for many countries (but not the United States) during World War I, when many timber-importing countries experienced timber shortages because of disruptions to the North Atlantic timber trade.

Political elites in softwood-deficient nations affected by World War I instigated the first large-scale state-sponsored afforestation programs to augment their limited timber supply. In Britain, an island nation without sufficient forest cover, the wartime and postwar softwood timber shortage encouraged the government to subsidize and plant large-scale softwood plantation. On the recommendation of the 1917 Acland Committee, which inquired into the causes of Britain's timber shortage during the end of World War I, the Lloyd George government passed the 1919 Forestry Act and created the United Kingdom's Forestry Commission in order to create new timber supplies.[24] The Forestry Commission oversaw a massive softwood afforestation program on the UK's millions of acres of heath, grass, and mountains that were deemed at the time by foresters and politicians as "waste lands."[25] Other softwood-deficient countries, such as Australia, South Africa, and New Zealand, used the wartime scare as an impetus to push for the establishment state softwood plantations in the 1920s and 1930s.

Yet it was not so easy to successfully grow trees in plantations. To create timber plantations foresters had to overcome a series of environmental, scientific, and economic challenges. First, foresters faced the problem of selecting exotic species for a variety of climates and soils. European foresters had few practical suggestions for how to select and experiment with exotic species, so this knowledge had to be learned from scratch. Second, foresters had to break with European traditions by devising new, unorthodox silvicultural theories and methods. European methods and techniques for planting and managing softwood plantations were applicable only to a few indigenous European species in select temperate climate conditions, and by the 1930s even these methods came into question. Third, forestry had to be placed more firmly within the capitalist market economy. European forest management plans emphasized sustainability in forests and revenue rather than running forestry as a business attuned to market and ecological cycles. Foresters studying the silviculture and economics of plantations began to challenge this tradition and instead tried to reorient forestry management to take into account of economic cycles and use the motivation of "profit and loss" to encourage productivity gains.

Foresters working outside Europe, if they lacked a useful indigenous species that was either abundant or easily grown, had to figure out how to select exotic trees with desired characteristics that would grow in local conditions. European silvicultural research provided little insight into this critical problem. Most foresters prior to the 1950s lacked an understanding of where species came from. There were inadequate supplies of seeds, and what seeds existed were often classified improperly. European foresters could offer little practical advice for how to select and test exotic species. In 1910, the South African forestry researcher C. C. Robertson noted: "In the other branches of science of Forestry we can look to some other countries, and particularly Germany . . . but the scientific naturalization of exotic trees has so far received comparatively little attention in these countries."[26]

State scientists and private individuals worldwide conducted thousands of experimental trials of varying quality with exotics. By the 1920s, foresters could more confidently plant exotic species because results from experimental trials started in the 1880s to 1900s were beginning to produce results. By then it was clear that foresters could plant Monterey pine (*Pinus radiata*) in New Zealand, Chile, and most of southeastern

Australia; *Pinus patula*, *Eucalyptus grandis*, black wattle (*Acacia mearnsii*), and blackwood (*Acacia melanoxylyn*) in South Africa; loblolly pine (*Pinus taeda*) and slash pine (*Pinus elliottii*) in the American South; Douglas fir (*Pseudotsuga menziesii*) in the U.S. Pacific Northwest, Britain, New Zealand, and Argentina; Japanese larch (*Larix kaempferi*), Sitka spruce (*Picea sitchensis*), and Scots pine (*Pinus sylvestris*) in Britain; *Eucalyptus sp* in Brazil; and teak (*Tectona grandis*) in India, Burma, Indonesia, and Trinidad.

By the 1920s and 1930s, some foresters began to believe that European methods for establishing and managing plantations lowered the potential growth rates of exotic trees in South Australia and South Africa.[27] This problem was attributed partially to how Europeans spaced and thinned trees in plantations. European silviculturalists recommended closer spacing and lighter thinning of trees.[28] When they did thin trees, European foresters thinned a plantation subjectively by thinning based on what they thought it should "look" like.[29]

The South African forester Ian J. Craib controversially challenged this wisdom. Craib demonstrated in various studies from the 1920s to the 1940s that wattles and pines could be grown more quickly if they were spaced more widely and underwent prescribed heavily thinning and pruning to reduce the competition for soil moisture and light among trees. The Forestry Division in South Africa took the bold move in the early 1930s to manage all of the country's pine plantations by spacing them widely and thinning and pruning them heavily based upon Craib's methods. This decision was a big gamble—if these methods produced trees with inferior timber, the country would lose millions of pounds and decades of work.

South Africa's more intensive method of plantation silviculture was, along with a growing recognition of the importance of genetics (discussed in the next section), one of the most debated topics in forestry from the mid-1930s to the early

1950s.[30] South Africa's more intensive style of exotic timber silviculture faced extensive criticism from European-trained foresters who challenged it on four fronts.[31] First, cautious foresters warned against embarking on a widespread experiment that might lead to ecological and economic failure. They argued that it would be better to test the theory and apply it to second-generation plantations rather than first-generation ones. Second, growing trees more quickly also raised intuitive problems that would take time to answer. Fast-growing plantations could potentially create a "second rotation decline problem" whereby fast-grown trees would deplete the soil's nutrients, leaving it deficient so that it would "be incapable of carrying a second crop of the same species."[32] Third, foresters worried that fast-growing wood from fast-growing trees would not be as useful or marketable as timber from slower-growing trees because they would be less structurally sound.[33] Fourth, South African plantations were seen as being almost tantamount to agriculture because of their intensity and emphasis on timber production.

Ultimately, the criticisms raised against South Africa's more intense methods of silviculture proved unfounded. South Africa's heavily thinned, wider spaced, weeded, and genetically selected timber plantations became the model for many timber plantations created for roundwood (i.e., the complete trunk, which could be used whole or sawn into boards) around the world from the 1940s to the late 1980s. Foresters in South Africa, India, Kenya, New Zealand, Queensland (Australia), Britain, and the United States South followed Craib's methods or developed similar insights.[34] South African methods of using wider spacing and heavier thinning were validated by two external reviews by the British forestry economist W. E. Hiley in 1947 and 1957.[35] In the end, they produced healthy, profitable pine plantations. Craib also correctly predicted that the rate of growth would not affect the strength of timber—experimental results from South Africa demonstrated this.[36]

The "second rotation decline problem," despite some initial alarms in spruce plantations in central Europe and pine plantations in Swaziland and Australia in the 1950s and 1960s, has not proved to be a serious problem globally, though there are now questions about whether this issue applies to intensive short-rotation tropical plantations.[37]

Some foresters wanted to subject forestry to market-oriented economics and business models. Pushed to its logical end (as it was in the second half of the century), forestry focused almost exclusively on growing and utilizing trees and selling timber products. Few foresters more ardently advocated subjecting forestry to market forces than British forestry economist W. E. Hiley. He put cutting-edge silvicultural techniques and market economics together to critique traditional European forestry theory and practice. He believed that European orthodoxies, such as the policy of thinning trees lightly and by subjective selection, gave rise to "the general belief that forests could only earn a very low rate of interest on their capital value."[38] This pervasive view discouraged landowners in countries like Britain from investing in timber plantations. Hiley made a daring move by quitting his job as a lecturer in forestry economics at Oxford University in 1932 to take up the appointment as the head of the Dartington Estate forests in Devonshire to prove that plantation forestry could be profitable for British landowners. When Hiley took the job, he told the owners that his first goal was to ensure that "the woodlands should be managed so as to yield the greatest possible profit or the least possible loss."[39] Hiley's venture did prove profitable. It signaled that private investors could finally begin making profits on plantations.

After solving silvicultural challenges and subjecting forestry to market forces, a number of technological problems still had to be overcome. Many of the world's most extensive forests could not be productively used. Second-growth pine forests in the American South and eucalyptus forests in Australia had beguiled investors and foresters for decades because they grew prolifically but the wood presented technical problems that made it difficult to use them industrially. A series of technical innovations in the 1920s made it possible for industries to utilize previously unusable resinous pines and hardwood eucalyptus. These innovations paved the way for future plantations. This set the stage for the rapid growth of plantations of subtropical pines and eucalypts. Today *Pinus* (the genus of pines) and *Eucalyptus* are the two most widely grown and most productive genera grown globally in plantations.

Old-growth pines in the American South had high levels of resin, which clogged up industrial processing and made them unusable for paper. The newspaper industry in America sourced its pines from the North, where they had less resin than older trees, and they imported timber from the Baltics. Working for the Georgia State Forestry Service's Paper and Pulp Division, the chemist Charles Herty discovered in the 1920s and 1930s that young second-growth pines did not have high levels of resin, and could be turned into white paper that was suitable for newspaper and printing, the largest market for paper at the time.[40] This insight allowed the American South's vast second-growth pine forests to become the natural base of a vast American timber products industry.

Another technical innovation happened when industry found a way to utilize eucalypts in papermaking. Prior to the 1920s, the shorter, stockier fibers of eucalyptus made it difficult to use industrially. In the 1920s, researchers and paper companies in Australia (as well as Brazil and South Africa) discovered how to make cardboard and strong, brown "Kraft" paper from eucalyptus fibers.[41] At the time, these innovations allowed Australian companies to finally begin utilizing Australia's extensive eucalyptus forests. It also meant that Brazilian and South African companies could use their plantations of eucalypts for more than just fuelwood, railway ties (sleepers) and mining props. These innovations provided the green light for the rapid global expansion of eucalypt plantations in the second half of the twentieth century.

Despite increased knowledge of how to grow certain tree species in select parts of the world (especially temperate and select subtropical climates), timber plantations still were risky and expensive projects. Many countries, especially those in the tropics, still had not been able to discover species that could be grown successfully and without financial loss. Plantations were also expensive. Mechanization had only just started in highly developed forestry regions in the 1920s. Foresters relied largely upon manual labor for planting, thinning, and harvesting. Foresters required a cheap, abundant labor force, something that was difficult to find and maintain.[42] In countries with high labor costs, such as Australia, forestry departments frequently used penal labor to plant and manage plantations.[43] South Africa established forest villages for poor whites to create and manage plantations during the 1920s to 1930s, as well as contracting migrant black labor. The U.S. government harnessed the vast labor force of the Civilian Conservation Corps during the Great Depression to plant 2.3 billion trees—half of all trees *ever* planted in the United States.[44] Such a massive effort has never been seen again in the United States because of the labor cost.

The growth of plantations stopped in 1939 with the outbreak of World War II. Countries that imported high volumes of timber before the war, including Britain, South Africa, and Australia, found themselves cut off from timber markets during the war because of naval blockades and the militarization of ocean shipping. Imports of timber into Britain fell by more than half from 1,207.7 million cubic feet from 1934 to 1938 to 477.0 in 1945. Australia's imports declined from 60.4 to 30.2 in the same period.[45] The United States sent some strategic timber in convoys to Britain, although 75 percent of timber used in Britain came from domestic forests.[46] Roughly 90 percent of the wood cut in Britain during the war came from private forests.[47]

When the war ended, timber plantations were seen as a necessary part of solving postwar timber shortages. Reconstruction required large amounts of softwoods for housing and construction. Europe faced a severe housing shortage as the result of the destruction of houses during bombing and artillery campaigns. Timber shortages slowed down recovery in Europe and demand for wood outstripped supply until 1948.[48] Timber shortages also existed in Australia, New Zealand, and South Africa during the late 1940s, hindering the construction of housing for returning soldiers and new homeowners. In 1951, Japan passed a new Forest Law that encouraged massive postwar tree planting from the mid-1950s to the 1970s.[49] These plantations now constitute approximately 40 percent of Japan's forests.

Postwar planners feared that domestic and global timber shortages would last long after the war's end. Foresters working for the newly created United Nation's Food and Agricultural Organization (FAO) stated simply in its 1948 survey of the world's forests that "the whole world is suffering from shortages of forest products." At the time, the forestry division at FAO estimated the world's forest estate to be approximately 4 billion ha.[50] There were only 8 million hectares of plantations globally (excluding Europe's seminatural forests), or less than 0.2 percent of the world's total forest cover.[51]

Plantations for Profit

In 1941, the Weyerhaeuser Timber Company opened America's first "tree farm" at its 56,000 ha Douglas fir plantation in Clemons, Washington. In its national advertising campaign it proclaimed proudly, "Timber is a crop." The tree farm was part of an advertising campaign that showcased the company's efforts to replant trees in native forests that had been previously cut. Weyerhaeuser's publicity branch promoted the idea of the tree farm to counter the widely held belief that large American timber companies merely cut existing forests but did not replant them. Weyerhaeuser also used the campaign to educate surrounding residents about the dangers of fire. Weyerhaeuser's internal economic forecasting had suggested that accidentally lit fires could wipe out the profitability of plantations. They used advertising and propaganda—including a popular Technicolor film *Trees and Homes* (1941) that Metro-Goldwyn-Mayer produced—and tours of the grounds by locals to stop neighboring residents from accidentally starting fires.

The slogan "timber is a crop" reflected a fundamental shift in the way developed societies produced timber during the second half of the twentieth century. Timber production transitioned slowly from being an important means to an end—for example, conservation—to becoming its own end. Timber plantations became more intensive, agricultural in method, and privately owned and operated. Whereas private industry planted only around 3,000 hectares of plantations per year before 1950, this figure reached 485,000 hectares (1.2 million acres) per year by the 1980s.[52] Today, private forests produce 89 percent of the timber harvested in the United States.

The rise of industrial plantations in the United States corresponded with the growth of the American Tree Farm System, which spread swiftly throughout the country following Weyerhaeuser's first plantation in 1941. In response to Weyerhaeuser's actions, the National Lumber Manufacturers Association created the "American Tree Farm System" in late 1941 to certify that private owners complied with its regulations. Forest owners quickly signed up to have their forests listed as American Tree Farms. The American Tree Farm System grew rapidly after the end of World War II. In late 1945 it included nearly 11 million acres, with private plantations in states in the South and West dominating the listings.[53] By early 1949, a total of 17 million acres of forests were certified. By 1984, the system included approximately 55,000 tree farms covering nearly 87 million acres. Corporate tree farms accounted for 75 percent of the 87 million acres.

The epicenter of plantations was located in the American South, a region that currently is, along with Brazil, the world's largest by volume producer of wood products. This postwar trend reflected a major reversal of the South's history. Until the late nineteenth century, the South's vast pine and hardwood forests remained a hindrance to the expansion of farms and ranches. For instance, old-growth longleaf pine forests once stretched along the coastal areas from the Carolinas to eastern Texas, covering approximately 60 million acres.[54] As late as the 1860s, most of these southern pine forests had not been logged.[55] The late nineteenth-century introduction of logging railways, which facilitated logging and the transport of timber, led to an intense period of rapid clear-felling. By the Great Depression, most of the South's old-growth forests had been cut, and crops, scrub, and young second-growth forests, comprised mainly of fast-growing pines, grew in their place. Most of these cutover forests were located on privately owned land.

Whether the American South could produce profitable pine plantations remained a serious question well into the twentieth century. Longleaf pines took approximately a century to grow to maturity, and slash and loblolly pines had wood with exceedingly high levels of resin that clogged up machinery. Charles Herty's (aforementioned) discovery in Georgia during the 1920s and 1930s that second-growth pines had less resin and could therefore be utilized in the Kraft pulp process raised the possibility that the South's second-growth pines could provide enough timber to supply a major industrial base. The South's warm and moist climate, its fast-growing native pines, good transportation system, extensive privately owned land, and lower wages made it an ideal location for a timber plantation industry to arise.

Some foresters recognized that plantations, followed to their logical conclusion, would turn foresters into farmers. H. H. Chapman, an expert on the ecology of southern longleaf forests and a professor of forestry at Yale University, worried about the rise of commercial loblolly and slash pine plantations. Writing in an 1954 article, "Do We Want a 'Pulpwood' Economy for Our Southern National Forests?," Chapman pointed out that this system would ultimately undermine the larger goals of the conservation model by crowding out public interests:

> Many, if not the majority, of the large southern corporations, faced with the problem of supplying their huge investments in plants with adequate raw materials, are tending towards the management of their own lands on short rotations of up to 30 years, cutting the crop clean, and repeating the process by planting or from seed trees . . . But the comparative values of the crops by each method are heavily, lopsidedly weighted in favor of the method of thinning for pulpwood and producing as an end product a crop of high quality sawlogs . . . But when more pulpwood is the dominant urge, the drive is to produce the greatest yield of this single product from which the mills derive their profit. Carried to its logical conclusion, this policy would neglect all other economic and public interests in order to achieve this one goal.[56]

A few owners of large southern forests decided not to pursue this industrial management program, and instead sought to maintain longleaf pine forests using the pioneering methods of Herbert Stoddard. It is important to note that these holdouts represented a very small percentage of southern landowners.

After Hearty's discovery that second-growth pines could be profitably utilized, the next great advance in plantations was the development of breeding methods that led to increased productivity gains in "third-" and now "fourth-" generation Southern plantations. Purposeful tree breeding, part of a larger cooperative tree improvement movement, began in 1951 with the founding of the first cooperative breeding program that brought together Texas industries, Texas state foresters, and university researchers at Texas A&M University.[57] The young American forester Bruce Zobel took up the role as the head of the cooperative in 1951. Zobel's influence on American (and global) forestry breeding research cannot be overstated: his work in Texas and the majority of his career working at North Carolina State led to the formation of a pioneering and dynamic relationship between university and state forestry researchers and private landowners. Zobel laid down rules for selecting, collecting, and breeding trees with superior genetic attributes. Over time, foresters in the public and private sector continued to select and grow the most elite trees from each generation. With each mature generation, foresters then reselect elite trees to be grown in the next generation. It is estimated that as of 2006, over 50 percent of all forest trees being planted in the United States were descended from a cooperative breeding program.[58]

State forestry researchers in Brazil, Sweden, Finland, Japan, Australia, New Zealand, and South Africa developed a variety of techniques to produce higher yielding timber plantations that allowed for the formation of profitable plantations. A number of critical innovations led to increased yields in timber production. These included a better understanding of the use of fertilizer (especially in nutrient-deficient soils), the importance of mycorrhizal associations (i.e., bacteria and fungi associations with tree roots), the necessity of weeding, and the critical role of breeding. Most important, foresters recognized the importance of parentage, which determined a tree's genetic makeup and growth potentials. One of the most pressing questions of the 1940s and 1950s was whether or not tree variation could be easily controlled, and potentially even manipulated, through selecting elite parent stock to be planted and bred. Foresters in the United States and throughout the British Empire hotly debated this question.[59] Many older foresters were skeptical about genetic improvements while many younger and up-and-coming foresters believed in the ability of humans to understand, control, and even manipulate tree characteristics.

The impact of breeding and genetic improvements can most clearly be seen in the history of eucalypt plantations in Brazil's southern subtropical states where short-rotation eucalyptus plantations pushed the boundaries of productivity. Brazil's dynamic plantation industry was the result of a synergistic relationship between the state and private railway and steel companies. Private companies worked closely with state researchers; funded research into silviculture, breeding, and technological utilization; and led planting efforts. In particular, the three-decade- long cooperation between the São Paulo state forester and agronomist Edmundo Navarro de Andrade (1881–1941) and the Paulista Railway Company paved the way for the rapid intensification and expansion of eucalyptus plantations in Brazil later in the century. Andrade worked with the Paulista Railway Co. to select and plant eucalyptus species that could be used for railway sleepers and fuel. Navarro wrote and spoke prolifically about the potential of eucalyptus, and served as a key advocate of eucalyptus.[60] By the late 1950s eucalyptus plantations, mostly planted and controlled by Paulista Railway Co., reached approximately 1 million hectares and produced over 6.6 million cubic meters of wood.[61] Over time companies and researchers in southern Brazil developed a mechanical and intensive style of silviculture that produced short-rotation plantations of eucalyptus and pines.

Tax incentives and state subsidies further catalyzed Brazil's timber plantation industry. The Brazilian government passed Law 510 in 1966 (and passed another similar law in 1970) that provided financial incentives in order to kick-start the plantation industry. Large private companies entered into the plantation market as a result of these incentives. The company Aracruz Florestal, in particular, built up massive eucalyptus plantations and industrial processing plants in the states of Espirito Santo and Bahia. Aracruz Florestal used breeding and intensive management methods to produce superior trees grown in ideal conditions. From its origins in the late 1960s, Aracruz Florestal by the early 1980s had built up over $100 million in foreign export earnings.[62] By the late 1980s, Brazil had approximately 6 million hectares of eucalyptus and pine plantations, mostly located in its southern states. These plantations supplied industries and provided a large amount of Brazil's foreign exchange surplus.

Chile followed a somewhat different path to building up a large plantation and wood products sector.[63] Initial state involvement and subsidies encouraged the formation of a robust private sector in the temperate south-central region of the country. The Chilean government with the help of private industry had planted 300,000 hectares of *Pinus radiata* plantations by 1974. That year Pinochet revolutionized the sector by passing the 1974 Forest Development Law, which secured private property ownership, privatized existing government plantations, and offered subsidies to industry to grow and process timber. These reforms instigated a rapid buildup in the size of plantations and hastened the development of industrial processing facilities in Chile. Post-Pinochet Chile opened itself up to foreign investment, which led to significant collaborations between Chilean and foreign firms.

Private growers entered the market more slowly in countries like Australia and New Zealand, where state and national governments subsidized planting programs, research, and infrastructure. Governments in the Antipodes viewed plantations as a "public good," reflecting the timber scarcities of World Wars I and II and the lack of abundant native softwoods. In each case, both governments subsidized the formation of state plantations and rejected free trade during the 1940s to 1970s as an economic and strategic means of supplying national timber resources. New Zealand's reliance on state subsidized plantations changed drastically in the 1980s when the government decided to begin privatizing them in 1987 and rely on private sector investment, management, and decision making. Australia, drawing on New Zealand's actions, began the processing of privatizing forests and processing plants in the late 1980s. [64]

Plantations in developed countries underwent extensive changes as a result of shifting economic conditions and political actions. The introduction of neoliberal economic and political policies in developed countries ushered in a period of privatization and government austerity in forestry development projects. In part, the state could act in this fashion because the growing success rate of plantations (domestically and globally) meant that governments no longer felt the need to subsidize them for strategic purposes. Capitalism and free trade, not strategy and protectionism, would dictate plantation policies in most developed countries. The full implications of this trend would be realized in the next two decades with the rise of China and the rapid growth in global trade.

Tropical Plantations: Exotic Reforestation

Eucalypts were the great hope of foresters in the tropics from the mid-nineteenth century onward. Because they were fast-growing and (supposedly) had antimalarial qualities and strong

Figure 2.3

Removing timber in India with an elephant, 1901

Photo courtesy of the Forest History Society, Durham, N.C.

timbers, species of eucalyptus received intense interest. In 1896 the great poet of Britain's Indian Empire, Rudyard Kipling, praised the heroic efforts of British foresters in India who, "experiment with battalions of foreign trees, and coax the blue gum [*Eucalyptus globulus*] to take root and, perhaps, dry up the canal fever."[65] Unfortunately, Indian foresters could not "coax" eucalyptus to grow everywhere. In most instances, the species that was planted withered in the hot Indian summer sun. Like other failures of acclimatization in the nineteenth century, Indian foresters planted eucalyptus species that came from cooler climates in warmer climates.[66] Poor species selection hindered efforts to make eucalyptus plantations in the tropics well into the twentieth century.

Planting native trees often proved just as difficult as acclimatizing exotics. Declining supplies of teak led to the first systematic efforts to make plantations from indigenous trees in South and Southeast Asia. There are many species of valuable tropical timbers, but no other tropical species received more attention in the nineteenth century than teak (*Tectona grandis*), which grows throughout tropical South and Southeast Asia. Foresters found it almost impossible to induce natural reproduction of teak in forests that had been harvested. After foresters cut down teak trees from a forest, hardier evergreen forests and shade-tolerant species grew up in their place. Berthold Ribbentrop, the German Inspector-General of Forests in India, lamented this fact: "We have evergreen, tropical forests, in which a luxuriant vegetation springs up as soon as the removal of the old forest growth admits sufficient light; but alas! it is more frequently not the vegetation we want."[67]

The seemingly obvious answer to India's teak shortage was to grow it in plantations. However, this proved extremely difficult to do. Every experiment in Bengal, Bombay, Burma, and Madras in the first half of the nineteenth century failed to produce viable plantations.[68] With the exception of the Nilambur plantation in southwest India, efforts to make teak plantations in the second half of the nineteenth century failed. The Nilambur plantation proved difficult to replicate because of high labor costs, different conditions (e.g., different regions were too hot, did not have enough rain, had poor soils, etc.), and teak trees, when planted together, suffered from higher rates of disease and pest attacks. Teak plantation failures continued into the early twentieth century, and the size of official government plantations actually declined after decades of attempts in the early 1900s after the Indian Forest Service abandoned some plantations.[69]

The inability to grow tropical plantations was a serious problem after World War II. Wood was required for the burgeoning population of the developing world, located primarily in the tropics. Western leaders recognized that sleeping Asian giants, such as India, had large, growing populations requiring ever more resources, especially timber products, vital for raising literacy rates and making an industrialized economy.

Tropical forests were difficult to utilize. Tropical forests featured a high diversity of tree species, most which had no known industrial application, and regenerating higher-value trees, such as teak, was difficult because they grew sporadically among other species. Western forestry advisors with experience in the tropics had by the 1960s became resigned to the fact that developing countries in Latin America, West Africa, and South and Southeast Asia could not supply their timber wants by sustainably cutting and regenerating indigenous tropical forests. Many international forestry advisors began to believe that sustainable yield models for forest management might not work in tropical forests.[70]

Tropical forestry experts advised developing countries to focus on selecting "superior" genetic strains and species (most of them exotic) and using "intensive, largely artificial practices for the regeneration of such species in single, uniform, or simple, early-successional species mixtures fitted to optimizing growth and yields."[71] This was an ambitious agenda considering that most tropical timber plantations, with a few exceptions such as teak in localized areas or exotics on higher elevation sites, had proved unsuccessful. John Phillips, a leading international advisor on tropical agriculture and forestry, noted bleakly that "the introduction of exotics into indigenous forest in low and medium elevations in the tropics, however, rarely if ever has succeeded except very locally."[72] The approximate size of tropical plantations in the 1950s was a paltry 0.68 million hectares, mostly in teak plantations in Indonesia.[73] FAO analysts warned in the 1963 *World Forest Inventory* that there is "insufficient evidence to assert that the present plantation programs are making serious headway against the destructive pressures exerted on the forest by a fast-growing population."[74]

To overcome this challenge, foresters worldwide instigated a major coordinated effort to find species that would grow in moist tropical conditions. Foresters worked with and for a variety of international forestry organizations to solve the enigma of tropical plantations. At various times, these organizations included FAO (UN), the U.S. Forest Service's International Forestry Division, the International Monetary Fund, the British Middle East Office, the World Bank, Australian AID, among others. Though forestry advisors saw their role as being one of helping to promote economic growth, and thus, creating better livelihoods, forestry aid served a dual purpose for many Western countries. The United States and Britain even used forestry as "foreign policy" to maintain influence in former colonies and perhaps even used foresters to gather intelligence (e.g., the CIA funded 20 percent of the budget of the International Forestry Division).[75]

Timber plantations were seen as a viable means of industrializing and modernizing. Economists and foresters believed that timber plantations produced an economic multiplier effect. The primary production of timber led to the development of secondary production and services. The strongest advocate of this economic theory was Jack Westoby, an energetic forestry economist who directed much of FAO's forestry agenda for the 1960s and 1970s. In a seminal paper published in 1962, Westoby argued that forestry served as a fundamental agent of economic modernization in developing countries because the production of wood brought in capital, supplied jobs, and created value-added industries (e.g., papermaking, etc.) that encouraged industrialization.[76]

Westoby's economic philosophy guided FAO technical advice to developing countries from the mid-1960s to the mid-1970s. FAO experts, who came from a variety of national backgrounds and forestry careers, designed experimental trials, provided technical literature, and worked with indigenous foresters to find solutions to local problems. Leaders from developing countries invited FAO forestry advisors into their country to advise on large-scale planting efforts that would stimulate industrialization, promote domestic economic growth, provide export income, and create infrastructure linking remote areas to metropolitan ones.

National and regional governments and state forestry departments dominated planting programs in developing tropical countries. Newly independent nation-states in West Africa, Southeast Asia, and South Asia controlled large state forest reserves inherited from colonial forestry legislation. The state controlled large swaths of reserved state forests, usually composed of a diverse mix of forests with only a few small exotic plantations of native and exotic trees. State foresters in Southeast Asia received powers from colonial legislation to strictly police forests, though, in reality, enforcement of forestry laws declined significantly from the colonial period, as loggers (illegal and legal) and government foresters pushed back forests to make smaller farms as well as larger plantation complexes.[77] In other places, such as India, state foresters saw local peoples as problems—they hindered the harvesting and planting of trees and were blamed as causes of illegal deforestation.[78] Forest reserves were eyed for their valuable hardwood timbers and as sites to create plantations of agricultural products. Governments viewed tropical forests as unproductive wasteland waiting to be selectively cut or converted into some other form of production.

Forestry experts advised developing countries to move away from the sustainable yield model of forest management they had inherited from European colonialism. Prior to decolonization, colonial forestry programs emphasized extensive management and natural regeneration (if possible) and only planted exotics on select sites. This ran contrary to what FAO forestry experts recommended from the 1960s onward. For instance, Eric Lundqvist, the FAO technical expert sent to India in the early 1960s, advised the Indian government to give up the British colonial conservation model they inherited and to instead grow more exotic timber. The report argued that to alleviate India's growing pressure for wood in the coming decades, "there will have to be a revolution in thought and practice, from the concept of a forest which as little as possible removed from a naturally occurring association, to one which is producing the highest possible yield per unit area. In place of low-yielding mixed forest, extensively managed, there will have to be concentrated pure plantations of fast-growing trees, mostly exotics."[79] FAO experts advised "replacement plantings," replacing existing diverse forests with a higher yielding exotic species such as *Eucalyptus*, *Gmelina*, *Terminalia*, *Cedrela*, *Acrocarpus*, *Pinus*, or *Acacia*. The 1967 meeting in Canberra, Australia, "FAO World Symposium on Man-Made Forests and Their Industrial Importance," reinforced this view.

Most timber plantations in the developing tropics during the last four decades of the century were replacement plantings. Replacement planting—logging native forests and then replanting with another tree or crop—was a major cause of global tropical deforestation before and after the 1980s.[80] From 1960 to 1980, timber plantations probably led to approximately 6–8 million hectares in native forest removals, primarily in South and Southeast Asia. But the science and politics of timber plantations could not easily be separated from broader agricultural developments involving agricultural trees used for small farming, state colonization schemes, and public and privately funded industrial plantations.[81] An FAO meeting of experts in 1967 noted that "the logic of the distinction between agricultural tree crops and forestry tree crops is often obscure. There seems no good reason, for example, why plantations of rubber trees are thought of as an agricultural crop, while plantations of tan-bark acacia trees are classed as a forestry crop."[82] If considered together, plantations composed of trees—palm oil, rubber, eucalyptus, teak, cocoa coconut, and so forth—accounted for up to half of deforestation in many tropical countries.

Replacement plantings destroyed native forests but often did not produce economically or ecologically viable plantations. Stocking rates, the percentage of canopy coverage in a planted area, remained low. A review of global plantations suggested that many stocking rates were as low as 10–20 percent after a few years.[83] The vast plantations reported by the Chinese government were not as impressive in reality as on paper: Chinese plantations suffered from chronically low success rates and yields, throwing the figures into doubt. For instance, the success rate of plantations between 1949 and 2003 was only 37 percent.[84] India, too, suffered from this problem. Indian eucalyptus plantations created during the 1960s produced only 36 percent of their expected yield.[85] Some deforested regions did not produce viable plantations or regenerate back into forests. For instance, in the Western Ghats, foresters destroyed rain forests in 1975 and replaced them with eucalypts, most of which died in the high rainfall areas.[86] Invasive species intermingled with the remaining eucalypts to create a novel ecosystem radically different than the previous forests.[87] These failures were caused by a complex set of factors, including poor site locations, improper methods, a lack of weeding, the impact of fire, and poor species selection.

One of the most significant tropical plantation failures involving replacement planting was the Brazilian Jari Florestal project, funded by the American billionaire Daniel Ludwig.[88] After purchasing roughly 1 million acres of forested land in 1976 in the lower Amazon basin bordering the Jari River, Ludwig had much of the virgin forest destroyed to make way for cattle ranches, agricultural land, and 100,000 hectares of timber plantations composed primarily of *Gmelina* and *Pinus*. Unlike in Brazil's south, where plantations were managed mechanically and intensively, the Jari plantations of gmelina were planted on former native forests and were less intensively managed.[89] The first attempts to create Gmelina plantations suffered a high failure rate, and Ludwig was forced to sell the project—after much criticism—to Brazilian investors in the early 1980s. Globally the Jari Florestal project came to symbolize the dangers inherent in replacement plantings.

Foresters began to worry about the negative impacts of forest-led industrialization. Westoby began to question his earlier advice. In 1973 Westoby asked his FAO forestry staff: "Are we right in encouraging (or assenting to) the liquidation of slow-growing mixed hardwood forests and replacing them by fast-growing plantations?"[90] Five years later he refuted his 1962 thesis on forestry and development at the Eighth World Forest Congress in Jakarta, Indonesia.[91] He argued that forestry and forest industries did little to create an economic multiplier effect in developing tropical countries. Profits from cutting down native forests and exporting lumber were rarely reinvested in the rural economy and instead went to elites in urban areas and foreign-owned corporations. In developing countries timber industries were supposedly creating a type of economic dependency that stagnated economies, keeping them reliant on foreign capital and markets, and pushing down wages and skilled job creation.

A "bottom-up" social perspective focusing on rural people gained popularity among international forestry circles in the late 1970s. In 1978, the World Forest Congress meeting focused on the theme "People and Forests." This conference solidified the development of two interrelated forms of forestry: social forestry and community forestry.[92] Both social forestry and community forestry focused on encouraging socioeconomically disadvantaged groups, such as poor tribes, women, and rural people to plant trees that would be managed communally and used for a variety of products. "Community forestry" referred to a broad array of activities, including more intensive forms of rural timber production to the management of small groves. "Social forestry" often focused more narrowly on alleviating fuel wood shortages by encouraging more local production of wood products. Both initiatives recognized and drew from the plethora of local forestry production systems already practiced by people in developing countries.[93]

The root of people-centered forest management dates back to well before the origins of scientific forest management. Dietrich Brandis unsuccessfully called on the British Indian government to institute a system of village forestry, involving the incorporation of local residents into management schemes, in the late 1860s to mid-1870s.[94] Brandis recognized that the implementation of state laws that took away access and control over traditional forests would alienate local residents, who would in turn take less interest in conserving forests or obeying laws. His astute observation was correct: conflicts over forest access and usage became an enduring result of colonial forestry policy. Postcolonial social and community forestry reflected growing frustration felt by communities living near forests who protested their inability to use forest resources and also pointed out the hypocrisy of state policies that were supposed to conserve forests but, in reality, allowed local officials and business interests to plunder public forests legally and illegally.

Many foresters listened to the concerns and needs of local residents when visiting developing countries, especially those in South Asia.

The 1970s emphasis on rural development brought tropical foresters even closer to agriculturalists. A variety of publications and conferences, ranging from the Jakarta Declaration at the 1978 World Forest Congress to the 1979 World Conference on Agrarian Reform and Rural Development in Rome to the World Bank's 1978 *Forestry Sector Policy Paper* emphasized the importance of forests in rural development and agriculture.[95] International foresters recognized that rural development required an integrated approach. A seminal FAO Forestry Department staff report in 1981 (based on a paper from the 1979 Rome conference) noted in the first sentences the traditional divisions between forestry and agriculture: "Until recently, rural development was generally considered from the point of view of food and agricultural production. Forestry tended to be regarded as a separate and isolated sector of interest only if it were possible to promote wood production for export or domestic industry or necessary to regulate water supply or control erosion. Attention was focused on the control of dense forests or on the creation of large plantations."[96] It went on to argue that these distinctions were becoming increasingly blurred in tropical forestry development. There was a growing "widespread recognition of the potential of forestry support for agriculture, especially as regards the small farmer."[97] The most important factor in this was the growing success of a select number of species, which could be grown in short rotations and used for a variety of purposes, including agro-forestry (trees and crops) and as fodder for animals.

Social forestry programs worked with varying degrees of success in rural areas around the world.[98] Impoverished rural areas presented many challenges. Frequently people in rural areas did not receive proper information about the planting, care, and uses of exotic species.[99] In India, the widespread planting of eucalyptus through state, social, and community forestry programs produced a highly politicized debate.[100] The "Great Eucalyptus Debate," as it became known, raised questions about the ecological and social problems caused and exacerbated by eucalyptus planting. Critics argued that eucalypts destroyed local biodiversity, used more water than other crops, crowded out crops that could be consumed locally, and lowered soil fertility. At the social and economic level, eucalypts took longer to grow than traditional agricultural crops, created higher rural unemployment, promoted absentee landlordism, and raised the price of grains.[101] In spite of problems, social forestry and community forestry programs blossomed in the 1980s and after. These programs have flourished because they directly engaged individuals and communities.

Timber plantations became more ingrained in the ecology and economy of tropical countries, especially in Asia, during the 1980s and afterward.[102] Three important factors shaped this trend. First, the effort by professional foresters to find trees that grew quickly in a variety of tropical conditions began to pay off. By the 1980s, foresters had found a variety of trees that could be grown successfully in a variety of tropical conditions.[103] With proper species selection and methods, it finally became possible to produce fast-growing industrial plantations on short rotations. New technological innovations meant these timbers could not only be used as firewood and roundwood but also turned into pulp, paper, veneer, plywood, and other wood-based products. The enigma of tropical forestry—at least silviculturally and technologically—had finally been overcome.

Second, the timber market became increasingly globalized from the 1960s onward as a result of the liberalization of trade policies on wood products, first in Japan and then in the West. This provided a growing market for indigenous and plantation wood exports. Japan, in particular, drove the terms of trade from the 1960s to the 1980s by importing half of the world's net timber trade during this period, three quarters of which came from Southeast Asia.[104] Plantation-grown timber did not figure as an important part of global timber trade during the 1960s to the 1980s, though it grew in importance from the 1990s on. Most timber exports, especially in Asia and West Africa, came from native forests. Stocks of native tropical hardwoods suffered severe declines in the late 1970s and 1980s because of increased rates of deforestation. This led to growing international recognition of tropical deforestation as a major global political and environmental issue.[105]

Prior to the mid-1980s, investment in timber plantations lagged behind overall development and private investment in agriculture, energy, and industry. Major development banks contributed less than 1 percent to the forestry sector during the early 1980s.[106] Japan, the primary purchaser of Southeast Asian tropical timber, did little to invest in plantations or in the reforestation of forests logged for sale to Japan.[107] Financing for plantation projects came primarily from government-run corporations and state forestry departments. Private industry initially shied away from investing in plantations in countries such as Indonesia because of the financial risk.[108] Increased state, international development, and private investment during the 1990s spurred a rapid phase of growth in timber plantations that continues to today.

Third, the intensification and expansion of tropical deforestation provided a justification in the 1980s for the creation of more and larger state and privately financed industrial timber plantations, primarily composed of exotic species. These plantations would feed a growing number of large industrial pulp and paper facilities that were capitalizing on the expanding timber trade and easier finance terms.[109] The Indonesian government implemented the Hutan Tanaman Industri (HTI) resettlement plan in 1984 to create plantations on cutover lands to supply massive export-oriented pulp mills. Indonesia's HTI settlements were supposed to stop deforestation and reforest degraded land, but they often encouraged illegal logging in native forests to make way for exotic timber plantations.

As a result of these dynamics, the expansion of industrial timber plantations in tropical Asia became increasingly divisive in the 1990s. Resistance to exotic plantations grew in developing countries as a result of growing public awareness of tropical deforestation, which became a topic of global concern in the 1980s, and the growth of environmentalism among poorer and rural peoples encouraged by international and national activism. Whereas in Western countries environmentalists succeeded in challenging clear-felling, environmental movements to stop the creation of plantations developing countries have been less successful, despite attempts such as the World Rainforest Movement's Montevideo Declaration in 1998 to raise global awareness about the negative social and ecological impacts of industrial plantations. Protests have been able to stop particular projects, but the financial and political interests of investors and elites; the global demand for timber-based products; and the massive profits to be made in plantations make it difficult for local resistance to change national policies.[110]

Success?

The global ascendency of timber plantations occurred because of internal developments within the profession of forestry and external demands from political elites and market forces. Foresters worked to rationalize timber production through silvicultural and technological advances. They succeeded. By the last decades of the twentieth century, foresters throughout the world knew which trees could be grown successfully in plantations and utilized industrially. These successes, with few exceptions, stemmed from state and university research programs. The world's timber is grown today with knowledge gained by over a century's worth of public financial support.

The success of plantations allowed governments to move away from the production side of timber and toward a role as a caretaker of forests. In the 1980s, a number of Western governments, influenced by the rise of neoliberalism and enticed by the possible profitability of selling or leasing plantations, started pulling funding away from silvicultural research and the creation of state plantations. The private sector, which gained much of its knowledge through state research and cooperation, began to take over the business of growing timber from the government. Private industry has not looked back, and governments around the world have decided to let private industries and free trade regulate the production of timber.

3

Native Forests: From Multiple-Use to Protected Areas

Rip rip woodchip—turn it into paper. Throw it in the bin, no news today. Nightmare, dreaming—can't you hear the screaming? Chainsaw, eye sore—more decay.

In 1990, the Australian country musician John Williamson released the song "Rip Rip Woodchip" to protest the clear-felling of native eucalyptus forests to make wood chips to export to the Japanese market. Williamson's song was part of a larger popular movement that called on Australian state governments to limit the use of clear-felling in state-controlled forests. Many feared that clear-felling on a large scale destroyed the biological and ecological integrity of Australia's remaining unharvested native forests.

The rapid growth of the wood-chipping industry from the early 1970s onward fueled public fear about clear-felling. Prior to the early 1970s, foresters considered large swaths of Australian state-controlled forests to be economically unviable. The rapid growth of the Japanese economy provided a new export market for large volumes of wood chips, which the Japanese used to create their own domestic pulp. In response to this new market, states throughout Australia opened up large state-controlled native forests to private timber companies that employed the clear-felling harvesting method to cut down all standing timber to be processed in mills located at ports. By 1990, Australia became the world's largest exporter of wood chips, a sizeable figure when one considers that forests only cover 6 percent of Australia's landmass.[1]

Australian foresters did not anticipate that clear-felling would cause a negative uproar. They justified clear-felling based on ecological and economic grounds. They sometimes argued that cutting down all standing timber and then burning the debris mimicked the effects of fire, a natural disturbance in Australian forests. Industry advocates proposed that it was economically unviable to selectively log individual trees given the low value placed on the timber of many species of eucalyptus. Wood-chipping created jobs in rural areas that had few other forms of employment. One of the reasons why Williamson received criticism from the timber industry was that it was assumed that he would, as a country singer, support rural jobs.

Australia was not the only country to undergo extensive public debates over how state forests should be managed. Debates about the ecological and economic impact of clear-felling were particularly intense in the United States, New Zealand, India, and Canada, among other countries, from the 1970s to the 1990s. During this period, public advocacy groups and environmental scientists questioned the authority of professional foresters to manage public forests. Critics argued that foresters focused narrowly on timber extraction at the expense of other conservation concerns. These claims challenged the scientific and moral authority of foresters, whose expertise and power rested on their longstanding assertion that they acted in the interest of the public good. The controversy over clear-felling and the declining prestige of foresters led environmental advocates to call for the creation of protected areas that limited or banned harvesting altogether.

Wars in the Woods

What these public and scientific debates shared was that participants all appealed to scientific, moral, and public conceptions of what nature *ought to be* to justify management decisions. "Nature," as Raymond Williams and William Cronon have famously noted, is one of the most complex words in the English language.[2] Conflicts in the 1960s and after occurred because the emergence of new environmental ethics and scientific perspectives challenged dominant assumptions about forest management. Despite claims by various scientific bodies and public organizations, there rarely has been a "right" or "wrong" way to manage forests because there is no consensus view on what nature *is* or *ought to be*.[3]

Foresters were caught off guard by the intensity of public criticisms because they imagined themselves to be the true environmental stewards of forests. They felt that they were supremely suited to understand forests because they studied them and worked with them for their living. They generally resisted attempts by other professions and scientific disciplines to determine and implement management plans because they had confidence in their own professional abilities. Foresters were able to maintain this superior attitude for almost a century.

During the first half of the twentieth century, foresters maintained their control over government forest policy, although the period saw the development of new scientific disciplines that produced findings that empirically and theoretically challenged the assumptions underpinning forest management. Researchers in the fields of ecology, hydrology, and wildlife biology publicly and privately challenged key pillars of forest policy, such as the belief that forest cover aided water retention or that fire should be excluded entirely from forest ecosystems. Heterodox foresters also raised questions about the impact of forest management practices that limited natural diversity, consumed large amounts of water, and introduced potentially invasive exotic trees.

The post–World War II period witnessed significant shifts in public attitudes toward state regulation of the environment. Fears about human survival in an uncertain nuclear and chemical age also encouraged people to see that plants, animals, and entire ecosystems were at risk from human action. Rachel Carson's best-selling *Silent Spring* (1962) popularized fears of animal extinction by warning that the pesticide DDT was devastating bird populations in the eastern United States. The field of conservation biology, which was devoted to preserving the diversity of nature ("biological diversity"), arose during the late 1970s and early 1980s against this backdrop. By the 1980s, scientists in multiple fields warned that deforestation, ecological fragmentation, and human action threatened species globally. Conservation biologists saw protected areas and reserves as a key tool to saving endangered species and entire ecosystems. The prominent Australian ecologist David Lindenmayer writes, "It is perhaps not surprising that reserves have been the primary focus of conservation biologists since the discipline began."[4]

Protection advocates found a ready-made precedent—national parks—that justified the setting aside of land to preserve natural values. They expanded the concept of national parks and devised new terminology, such as "protected areas," which overlapped with national parks and influenced nature protection globally. The international heritage conservation framework established by the United Nations Educational, Scientific and Cultural Organization (UNESCO) 1972 World Heritage Convention further encouraged countries to conserve the world's "superb natural and scenic areas" by limiting or stopping intensive extraction of resources. There was a substantial increase in the number of national parks and protected areas globally from the 1960s onward as a result of public and scientific interest in protecting large ecosystems.

Some of the most politically significant land set-asides centered on government-owned forests that had been slated for harvesting. Foresters throughout the world believed that they could apply more intensive management practices, especially clear-felling, while continuing to conserve other nontimber values. Paul Hirt describes this as a "conspiracy of optimism" because foresters and government officials overlooked the downside of extractive policies.[5] The increase in extraction worried scientists and environmentalists who argued that extraction was destroying ecological integrity and aesthetic values and increasing the risk of extinction. A few of the most important events included the conflict over the establishment of Redwood National Park and the battle over harvesting in the Southwestern Wilderness in Tasmania. By the 1980s, state foresters faced increasing public pressure to limit, or stop altogether, logging in native forests. Public pressure and a series of political and legal decisions led to the disbanding of long-standing forestry institutions and declining harvests from state-owned forests in Western countries, such as the United States, Australia, and New Zealand. Timber production from plantations helped justify the setting aside of native forests. Countries as diverse as New Zealand and Thailand decided to rely upon domestic timber plantations and timber imports from foreign countries to offset declines in timber production from native forests.

Trust in Foresters

Determining who manages nature is a decision that requires a great deal of trust. This is particularly true of forests. Most native forests take over half a century or more to grow to economic maturity. By the dawn of the twentieth century, government officials and the middle classes in most developed Western countries believed that, out of all groups, foresters had the knowledge and ability to manage public forests. Foresters gained power by asserting the validity of their own professional views over competing claims from private landowners, farmers, forest residents, and other scientists.

Forestry arose as a profession at the same time that other technical professions legitimized their position in Western societies and governments during the nineteenth century. The historian of science Theodore Porter argues that the public in Europe and North America trusted professions that used quantification and technical language to justify professional decision making that was supposedly in the public's interest. Experts used statistics to present their professional expertise as "objective," and thus more valid than perspectives not based on quantitative analysis.

Porter's book further argues that the "public role of quantification reflects social and political developments . . . [and] cannot be reduced to scientific and technological ones."[6] In particular, technical professions tried to recruit from the social elite because attracting upper-class candidates gave such professions greater social legitimacy among elites and the public.[7] Many of the first professionals were imbued with a confidence and power that came from their class background while at the same time they professed to hold special technical knowledge that made them the final arbiters in key decision-making processes, such as where a road or dam should be built, how much people should pay for insurance, or how a forest should be managed. Though professional scientists often worked for democratically elected governments, many of their decisions were not regularly reviewed, and when they were reviewed, elected officials relied on the technical knowledge of experts to frame the problem and outline solutions.

Foresters selected recruits from the upper and upper middle classes to establish the legitimacy of the profession. The

German forester Dietrich Brandis noted that in Germany and France forestry was "a calling followed by young men of the best families."[8] In Europe, France took training a forester as an elite to its highest art form. One of France's post-Revolutionary *Grand Ecoles* (great schools) was the *Ecole Nationale des Eaux et Forêts* (the French national forestry school), established in 1824 along the same lines as the *Ecole Polytechnique* (established in 1794).[9] European foresters tried to export this social system elsewhere in the world. Brandis appealed to British Indian administrators that 'it may be found, as a rule, advantageous to give preference . . . to young men of good family connections."[10] American foresters initially came from an elite background. The movement to establish state forestry succeeded partly because wealthy American landowners and the business elite agreed to work with the federal government. One of the "founders" of American forestry, Gifford Pinchot, was an extremely wealthy and well-connected person (he became the governor of Pennsylvania). Pinchot endowed Yale University to establish a forestry school that would produce an elite corps of American-trained foresters.

One of the reasons why foresters and other professions sought to recruit the "right sort" of person was that many technical decisions had to be made without full knowledge about what the consequences of these decisions would be; in these instances, government officials and elites wanted people who supposedly had strong moral character and a proper education, characteristics associated with the upper class. This is because, as Porter goes on to show, complex decision making could not easily be determined by a preset formula, even though professions derived power from their ability to quantify the world.

At the pinnacle of their power, foresters made big claims about the benefits derived from forestry that had little, if any, scientific evidence to back them up. A key contention that professional foresters made from the early nineteenth century through to the mid-twentieth century was that forests moderated extremes in climate, stopped erosion, regulated stream flow, and mitigated floods. This argument underpinned global efforts to gain control of private and community forests. It justified forestry practices such as expansive afforestation.

To make these claims, foresters amassed a war chest of anecdotes about climate change and hydrological regimes. Yet their claims rested almost entirely on assumptions rather than being derived from experiments or careful long-term monitoring. The link between forests and water protection was not maintained universally by all segments of society. Engineers criticized the hydrological ideas of foresters by arguing that forests actually decreased (rather than increased) stream flow and that forest effects on major flood events were negligible.[11] Farmers noted that many trees dried waterways instead of producing more rain or conserving water. In spite of early challenges, foresters gained ascendency in government circles because of the supposed benefits of forestry, as well as because elites recognized the economic importance of tree planting and forest conservation.

Internal Critique

The authority of foresters peaked during the late nineteenth to mid-twentieth centuries. Despite forestry's role as a favored environmental science, the beliefs of orthodox foresters faced criticism from a new generation of heterodox foresters, practitioners of other sciences, and some vocal segments of the public. Unlike public conflicts after World War II, debates before then occurred mostly (but not entirely) among professional scientists. The roots of many of these early conflicts lay in the establishment and expansion of new fields in environmental science. Some of the key advancements in the fields of ecology, wildlife biology, and hydrology challenged key assumptions of the conservation model. As the century progressed, foresters found themselves having to work with specialists whose views diverged from theirs.

Yet it would be a mistake to view foresters as a homogeneous group who failed to accommodate new scientific views and attitudes. In fact, the earliest, strongest, and most sustained criticisms of forestry came from foresters, particularly those in German-speaking countries. During the late nineteenth and early twentieth centuries, Germany not only had the oldest history of institutional scientific forestry, but was the world's leader in forestry education, forestry science, and forest utilization. This meant that Germany had the institutional and intellectual capacity, as well as a long history of tree planting and forest management, from which to make inferences.

Concerns about the biological failure and unattractive aesthetics of plantations appeared in the second half of the nineteenth century. These apprehensions led to the development of the "back-to-nature" movement. Michael Imort describes back-to-nature foresters as those seeking to "return to uneven-aged, mixed stands, as well as a balance between the exigencies of timber production and ecological and aesthetic considerations."[12] Some of the key thinkers included Karl Gayer (1822–1907) and Heinrich von Salisch (1846–1920). Gayer, a Bavarian forestry professor, expressed concerns that clear-felling and selecting species without reference to climate made plantations more prone to disease and destruction.[13] He proposed more natural silvicultural systems that took into account preexisting natural conditions of sites. The Polish-German von Salisch devoted much of his attention to solving the "un-natural" aesthetics of early plantations. His book *Forstästhetik* (Forestry Aesthetics) argued that beauty and recreation should be important considerations when planting forests and managing them.[14]

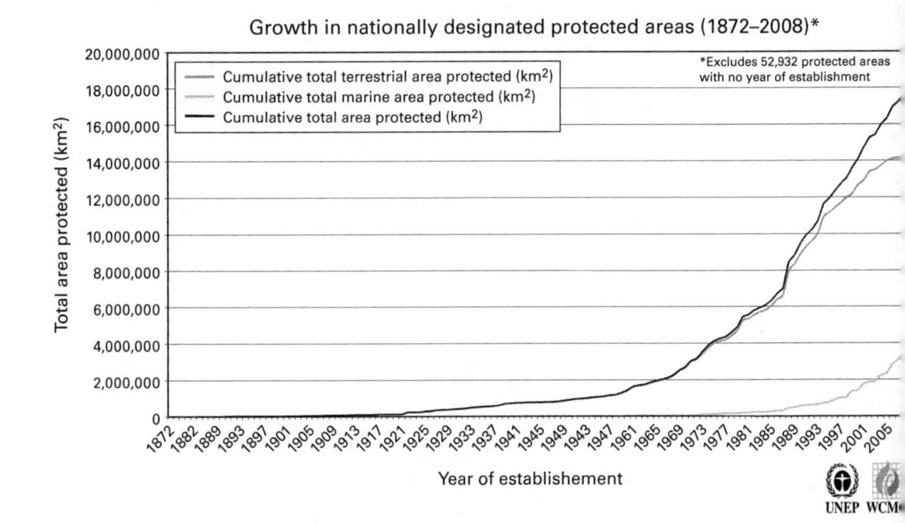

Figure 3.1

Historical growth of protected areas, 1872–2008

Credit: UNEP-WCMC, World Database on Protected Areas (Cambridge: UNEP-WCMC, 2009).

The movement to understand and replicate natural processes in order to produce healthier forests gained traction in the late nineteenth century. Adolphe Parade, a French forester, counseled his fellow foresters to "*imiter la nature et hâter son oeuvre*" (imitate nature, hasten its work).[15] To understand nature in a wider sense meant studying the interrelationships and life histories of plant communities, what the German Ernst Haeckel defined in 1866 as "*Ökologie*," or ecology. Yet, the development of the discipline of ecology in the late nineteenth and early twentieth centuries did not initially lead to a major reimagining of how foresters saw forests.

Ecological thinking opened the door for foresters who argued that forests should be thought of as holistic systems.

This holistic thinking strongly influenced influential northern European foresters. Two of the most important adherents were the German forestry professor Alfred Möller and the Finnish nationalist and politician A. K. Cajander. In 1909, Cajander proposed that there were distinct "forest types" that could be defined based on "indicator species" of ground flora.[16] His model went so far as to downplay trees as the basis for classifying forest ecosystems. Canjander's nationalist-inspired research argued that there were distinct ground floras that distinguished Finland's forests from those in Russia as a way to justify Finnish separation from Russia's control.

The concept of Dauerwald, or "permanent forest," was first advocated in 1920 by the forestry professor Alfred Möller. Dauerwald as a philosophy called on foresters to think of forests as a broader "organism," including all species and interrelationships, rather than seeing them merely as an aggregate of timber-producing trees.[17] Dauerwald called for selective cutting rather than clear-felling and rejected monocultures in favor of diverse forests relying on indigenous species that were adapted to sites. Möller wanted to encourage natural reproduction of forests to ensure the sustainability of forest productivity and ecological health. Möller's views sat on the intellectual edge of forestry orthodoxy at the time, and, though they received early positive attention, soon fell out of favor.

The concept of Dauerwald became the basis of official German forest policy for a brief period under the Nazi Party in the mid-1930s.[18] Hermann Wilhelm Göring, the second most powerful person behind Hitler in the Nazi Party, believed in the Dauerwald model. In early July 1934, Göring appointed Walter von Keudall, a landowner who practiced Dauerwald on his property, to direct the Reich Forestry Service [*Reichsforstamt*]. In 1935, Göring introduced the Reich Conservation Law, a sweeping law that took conservation powers away from the states and centralized it with the federal government. With new centralized powers, von Keudall demanded that landowners protect ancient trees, clear-fell less than 2.5 percent of their forests, and cut the worst (not the best) trees to improve the genetic health of forests; the rules prohibited the cutting of conifers less than fifty years old.[19]

Dauerwald did not achieve its idealistic vision. Frank Uekötter points out that forestry, like other environmental policies in Nazi Germany, was subject to "violent swings" based upon the fluctuating standing of individuals and groups who directed policies.[20] Many foresters and landowners disliked the program from the start. It lasted only three years before a watered-down successor policy based on "natural forest economics" [*Naturgemäßer Wirtschaftswald*] was introduced in 1937 to allow for clear-felling and higher extraction rates. Despite failing to break Germany away from its traditional forest history, the brief implementation of Dauerwald encouraged foresters elsewhere in the world to believe that it was possible to implement policies to maintain ecological health of wider forest communities.[21]

The move toward a more natural system of forestry in Germany had parallels in the United States. Three agencies, the U.S. Forest Service (USFS), the Bureau of Land Management, and the National Parks Service (NPS) controlled the vast majority of federal forests. Though the USFS, which is located within the U.S. Department of Agriculture, at various times attempted to take over the governance of national parks, the two have always been separate since the National Parks Service was founded within the Department of the Interior in 1916. NPS controlled vast forested tracts in famous parks such as Yellowstone National Park, Glacier National Park, and Yosemite National Park. At first there was little difference in scientific opinion about parks management, but divergences appeared from the 1920s to the 1940s when some USFS and NPS researchers started to question policies set by foresters.

Figure 3.2

Ranger on Cabinet Mountain National Forest, Montana, 1909

Credit: U.S. Forest Service photo courtesy of the Forest History Society, Durham, N.C.

Heterodox foresters and wildlife biologists within the USFS began to advocate for the maintenance of forests in their "natural state" for preservationist and scientific purposes. Aldo Leopold and Arthur Carhart, key forestry figures in this movement, pushed for the creation of vast areas within USFS-controlled forests to "be withheld and retained in as near a natural state as possible."[22] Chief Forester William B. Greely ordered the establishment of the first wilderness site (first proposed by Leopold) at the Gila National Forest in New Mexico in 1924. Designated wilderness areas were created to be at least 100,000 acres in total, a size imagined sufficient to allow for the self-regulation of plant and animal communities. By 1929, the USFS began establishing another type of untouched forest areas, research reserves, areas at least 10,000 acres in total. Unlike wilderness sites, these areas were oriented to facilitating forestry research. By 1945, approximately 1.5 million acres of USFS land were managed as wilderness. Prior to the passing of the 1964 Wilderness Act, the USFS on its own volition set aside 8 million acres of national forest as wilderness.

A similar turn toward natural management occurred within the National Parks Service. Early preservation management strategies in national parks in the United States and elsewhere in the world focused on a style of active management. In fire policy, NPS managers followed USFS guidelines, which saw fire as a destructive and generally unnatural disturbance. Park managers suppressed fires by building roads and fire lines through natural areas, removing dead trees and debris that could cause or exacerbate fires, and taking away burnt trees to allow for the regrowth of seedlings. Animal and insect populations were controlled actively by a variety of means, such as trapping, hunting, spraying, and culling.

The appointment of wildlife biologists to the scientific corps of the NPS introduced new scientific perspectives and generational attitudes toward park management that challenged orthodoxy. Some scientists began to express concerns that active management practices went against the preservationist ethic, which at its core sought to protect nature from human destruction. A popular ecological theory espoused by the adherents of Frederic Clements believed that, if human disturbance stopped, ecologies would move through various stages of succession toward a natural equilibrium, the "climax community." Nature is self-regulating, if only humans would stop interfering.

Whether or not to suppress fire remained one of the most pressing forest management questions in America for the rest of the century. In 1935, the wildlife biologist Adolph Murie (1899–1974) kick-started this policy debate by arguing against the instructions he received from a superior to remove dead trees from a large fire that struck a widely visited section of Glacier National Park. Murie argued that fire was a natural process and, without human intervention, the park would naturally have large sections of burnt-out forests. He challenged the orthodox assumption of foresters that large ecosystems needed active human intervention because their natural "equilibrium" had been disturbed. Lawrence Cook, the head of forestry operations in America's western national parks, was decidedly critical of Murie's argument. In response to Murie, he justified existing interventionist policies in a memorandum: "The parks have long since passed the time when nature can be left to itself to take care of the area. Man has already and will continue to affect the natural condition of the areas, and it is just as much a part of the Service Policy to provide for their enjoyment as it is to preserve the natural conditions. There is no longer any such thing as a balance of nature in our parks—man has modified it. We must carry on a policy of compensatory management of the areas."[23] Cook disagreed by arguing that human action had altered natural processes and, therefore, only humans could replicate "natural" processes.

Researchers elsewhere in the world—from India, Burma, and the American South to southeastern Australian and South Africa—debated the importance of fire as a naturally occurring phenomenon in forest ecosystems during the same period. Their interest was piqued by two problems relating to fire suppression and the regeneration of key species. First, in practice, it proved difficult to keep many of the world's ecosystems free of fire. Forest researchers noted these conditions especially in the ponderosa pine forests in the interior of the American West, eucalyptus-dominated forests in Australia, and chir pine forests of northern India. This fit a broader recognition led by ecologists that grasslands and certain heath ecosystems required fire for reproduction. Though fire suppression effectively lowered the incidence of fire across the twentieth century, it did not entirely stop fires from occurring, and it built up fuel loads that led to bigger fires in the future.

Second, foresters recognized that many commercially valuable tree species failed to regenerate after being harvested. Foresters found that less valuable species of trees and undesirable grasses and scrub often regenerated after they had felled valuable trees. Some foresters theorized that many trees required fires to initiate reproduction, either by releasing seeds, helping seeds germinate, or by reducing competition from shade-tolerant species that grew more slowly. By the 1920s and 1930s, some foresters openly rebelled against traditional European attitudes toward fire because they felt that orthodox principles worked against, rather than with, the natural conditions of forests and desired tree species. The British Indian forester, E. A. Grewswel, came to the conclusion that the management of Himalayan conifers (especially chir pine), was "based on pussy foot principles" that did not properly induce regeneration: "We talk glibly about following nature and forget that the nature we are visualizing may be an European nature inherited from our training and not an Indian nature."[24]

By the 1940s, it became less controversial to acknowledge that fire played an important role in many ecosystems and for certain species, but putting these policies into action remained highly controversial (see later this chapter). In America, managers in some parks, such as Yellowstone, began to instigate a series of limited trials in the 1950s to test controlled burns. Yet popular sentiment favored fire suppression.

A different set of circumstances led to conflicts about what constituted proper "natural" ecological management in South Africa. In South Africa most forestry efforts, with a few exceptions, focused on planting exotic trees. By 1935, plantation forests extended to about 300,000 ha and another 400,000 was predicted.[25] Foresters since the nineteenth century had championed tree planting as a means of producing timber, changing climates, improving streamflow, and beautifying forestless landscapes. During the first three decades of the twentieth century, South African foresters planted government plantations at the headwaters of rivers in mountain catchments in the southwest and northeastern parts of the country.

Attitudes toward exotic forests changed in South Africa as a result of tree planting. A growing number of white farmers and a handful of pioneering ecologists, amateur naturalists and high-level political officials started to believe that exotic pines, eucalyptus, and wattles used more water than did indigenous plants. Rather than conserving water, planted trees transpired water. Jan Smuts, a former prime minister of South Africa, expressed these beliefs at a conference of international foresters in 1935: "There is no doubt that a popular feeling is arising in South Africa that afforestation is causing the drying up of springs and water sources."[26]

Criticism of exotic plantations also stemmed from the rise of floral nativism, a movement that sought to celebrate indigenous plants and vegetation types and protect them from human destruction.[27] Initially, white residents saw the bareness of South Africa's landscape as something to be remedied with tree planting. Starting in the closing decades of the 1800s, whites in South Africa started to reimagine their relationships with nature. From once seeing forestless landscapes as ugly and unnatural, botanists and ecologists started to revel in the diversity, variation, and beauty of the plants. They sought to popularize this appreciation of indigenous plants through the establishment of indigenous botanical gardens and colorful publications celebrating South Africa's flora.

Criticism of exotic plantations came to the forefront of attention at the Empire Forestry Conference in 1935, held throughout South Africa. The meeting received extensive newspaper coverage, and was attended by leading politicians as well as foresters from around the British Empire. A special committee was devoted to questioning the hydrological impacts of exotic timber plantations. The discussion also included questions about the ecological basis of forestry in South Africa. Johann Keet, the chief forester of the Forestry Division, told the committee: "It is our plantations, especially, that stand suspect. They are accused of being ecologically foreigners to our climate, and South African foresters are accused of confusing natural forest conditions with exotic forest conditions, and that generally we have ignored the ecological outlook."[28]

At the core of the debate was the question of how to use insights from ecology to manage the environment. Keet defended foresters by noting that "the forester lives with nature, studies nature, and follows nature's law, and if that is not ecology, I fail to see what ecology can be."[29] Keet employed a functional view of ecology that focused on the tree instead of the forest. From a functional perspective, if a tree grew it was matched to the ecology because its growth suggested that it was receiving its biotic requirements. Forestry critics, such as John Phillips, disagreed with this view by emphasizing, along the lines of Möller and Cajander, that there were discrete, distinctly local "biotic communities." Exotic trees, for instance, supposedly changed the habits of bees and led to less regeneration of indigenous trees. This action changed thousands if not millions of years of evolutionary relationships.

There was one thing that ecologists and foresters at the conference could agree on—foresters needed to pursue empirical research on all the questions that were raised. That same year, the Forestry Division established a research station in the Jonkershoek Valley just outside Stellenbosch, located close to Cape Town. Research at Jonkershoek by the founding director, Christiaan L. Wicht, indeed confirmed what many critics had argued: exotic trees did use more water than many indigenous vegetation types.[30] At the same time, Wicht also recognized that all forms of vegetation used water. In his mind, the question of forestry was not *whether* to plant trees, but *where* to plant them. He argued for a holistic framework taking into account "ecosystems as whole" and the economic and ecological priorities of society.[31] Researchers from Jonkershoek devised a policy that directed afforestation to higher rainfall areas where forestry received higher economic returns than other land uses.[32]

Clear-Felling

No single issue undermined the environmental legitimacy of the conservation model in the United States, Australia, Canada, and New Zealand more than did clear-felling. Clear-felling involves cutting down all standing timber and either reseeding or allowing natural regeneration. The harvest rate increased in large government-owned forests in the United States, Australia, and Canada during the last three decades of the twentieth century. Extensive harvesting, often achieved through clear-felling, drew vociferous criticism from an increasingly coordinated antilogging campaign led by environmentalists. Conflicts over logging in old-growth and clear-felling put in motion a series of actions that ultimately restricted the power of foresters in Western countries.

In northern Europe, clear-felling had been a normal practice used to manage its planted and natural fir and pine forests since the mid-nineteenth century. Yet outside Europe, clear-felling often failed to achieve the same success. Regenerating

Figure 3.3

Firefighter at a 1936 fire in Santa Lucia National Park

Credit: USDA Soil Conservation Service photo courtesy of the Forest History Society, Durham, N.C.

valued species remained a thorny scientific problem. Clear-felling was utilized in the United States, Canada, and Australia before World War II, but to a much more limited extent than after. Experiments with clear-cutting in Douglas fir forests from the 1920s to the 1940s showed limited success.[33] Additionally, the extraction rate in USFS forests remained low as a result of stronger pro-conservation policies and higher extraction rates on private land. In Australia, two problems beguiled foresters. First, foresters still struggled with finding ways to regenerate more productive wet sclerophyll forests. Second, it was difficult industrially to turn eucalypts into paper and pulp.

Clear-felling came to the forefront of the forest industries in the United States, Australia, and Canada after World War II when foresters discovered that clear-felling was a profitable and efficient means of harvesting and regenerating redwoods in Northern California, mountain ash in southeast Australia, and Douglas fir in British Columbia, Washington State, and Oregon. Before the 1970s, most foresters expressed few concerns at cutting down old-growth trees, which they described as "over-mature" or "decadent" because they had already reached their economic maturity and would decline in size and health until they finally died and rotted. From an economic perspective, old-growth trees should be harvested because they perpetually decline in value. Forest policy in the 1960s and 1970s emphasized a policy of "liquidation-conversion" by raising the harvest rate to take advantage of the value of old-growths while replacing them with faster-growing second generation trees.[34]

The same biological capacities and ecological conditions that produced ideal timber-producing trees also formed forests that the public celebrated. By the 1870s, writers and tourists flocked to forests in Northern California and Victoria, Australia, to see giant trees and to experience the sense of being in nature's pantheon. Old-growth forests produce archetypal "sublime" forest experiences. Old-growth forests are often cool, shaded, and quiet, and they provide the ideal conditions for iconic ferns. Old-growth trees inspire a "cathedral" effect because they can grow to enormous heights (some upward of 100 meters or over 300 feet), have high biomass, produce straight timber, and be found growing as monocultures.

The tensions that erupted over these "sublime" forests arose after nearly a half century of consensus between preservationists and foresters. Preservationist leaders maintained cordial relations with state foresters from the late nineteenth century to the 1950s. A number of factors solidified this conservation

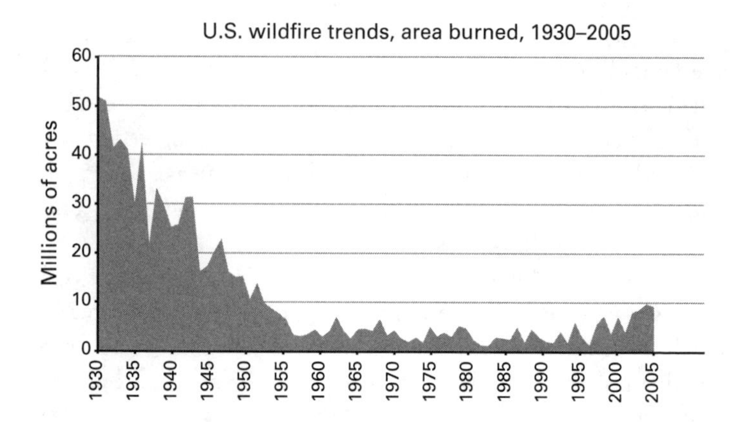

Figure 3.4

Area of United States burned, 1930–2005

Credit: Doug MacCleery, American Forests: A History of Resilience and Recovery (Durham, NC: Forest History Society, 2011), 36. http://www.foresthistory.org/Publications/Issues/American_Forests.pdf.

consensus. First, the early conservation and preservation movement in the United States, Canada, Australia, and New Zealand was dominated by a predominately Anglo elite who shared similar cultural values.[35] Anglo conservationists and preservationists worked together through consensus by discussing problems in private meetings in homes, clubs, and boardrooms. The Save the Redwoods League and the early Sierra Club fit this description. Second, most preservationists prior to the 1950s saw parks as one aspect of conservation; they did not challenge the utilitarian basis of conservation management in scientifically managed forests. Third, owing to economic and technical constraints and conservation policy, state forests throughout much of the United States and Australia had not been extensively harvested since before World War II. The rate of timber extraction increased rapidly from the 1950s to the 1970s.

Foresters encouraged conflict when they found ways to harvest and regenerate commercially valuable species. Researchers studied the biology of species to try to unlock their economic potential. Old-growth Douglas fir and mountain ash did not regenerate in narrow forest openings that were made after selective logging. In the Pacific Northwest, less valuable, shade-tolerant species such as western red cedar (*Thuja plicata*) and hemlock (*Tsuga heterophylla*) appear in openings after Douglas fir was cut. Researchers in the United States and Australia finally found solutions to these problems. In Portland, the USFS researcher Leo Isaac propounded clear-felling and artificial replanting as the most cost-effective and natural way of reproducing Douglas fir forests.[36] Isaac solved the riddle of regeneration that had confounded foresters, including those who called for clear-felling before the war.[37] The Australian forester Peter Attiwill advocated that clear-felling mimicked large-scale disturbance events that triggered the regeneration of mountain ash after fire.[38] New Zealand foresters recommended clear-felling for southern beach forests.[39]

In the 1950s and 1960s, forestry researchers advocated using clear-felling as a means to manage old-growth forests around the world. Foresters reasoned that if large-scale ecological disturbances (e.g., fire) and the biological dynamics of each species (e.g., fast growth after a fire and cohort senescence at advanced age) naturally produced less diverse forests, then foresters should seek to replicate large-scale disturbances by clear-felling.[40] Not only did clear-felling supposedly mimic the natural ecology of many forests, it cost less and was more economically efficient. Foresters were less concerned about the unsightly aftermath of clear-felling because they knew that trees would grow up to cover the bare ground with green again. Frederick Swanson writes, "The men of the modern Forest Service were not without feeling for the beauty of growing trees; in their eyes, the young plantations growing in clearcuts . . . were redeeming the forest from the destructive agents of fire, insects and decay."[41]

Postwar surges in global population and increased economic growth rates in Asia and the West led to a particularly steep increase in the harvest rate in government-owned forests in the United States, Canada (especially British Columbia), and Australia. In the United States, harvesting in state forests expanded by 5 percent per annum from 1945 to 1970, a rate that outpaced inflation and population growth. In the United States, harvest rates increased dramatically in the late 1960s and early 1970s in response to the rising price of plywood and the demand for new houses as part of Democratic president Lyndon B. Johnson's push for the "Great Society." By the late 1960s, clear-felling became the silvicultural method of choice in the American East and West.

The expansion of the harvest rate in American forests led to growing public concern about the cutting of public and private forests. These concerns corresponded with a rapid expansion in the number of public visitors who came to national forests and national parks for recreation and educational purposes. USFS propaganda was so successful in encouraging people to visit forests that it engendered a generation of Americans who wanted to protect the green forests they enjoyed from logging. It is one of the ironies in the history of forestry that professional foresters reached the pinnacle of their popularity only a decade before they began to face substantial criticisms from the public. A 1952 article in *Newsweek* noted that "most Congressmen would as soon abuse their own mothers as be unkind to the Forest Service" because of its sterling reputation.[42] By the end of the 1960s, this was no longer the case.

The USFS expanded the multiple-use system to include more nonindustrial forest usages, such as recreation areas and wilderness. This reflected national and international concerns among foresters and the public. Forestry attendees from around the world at the 1960 World Forest Congress in Seattle agreed that state forestry programs needed to take better account of the nonindustrial functions of forests.[43] That same year the U.S. Congress passed the Multiple-Use Sustained-Yield Act of 1960 in order to better assess the "relative values" of diverse forest resources. Though the concept and term "multiple-use" was itself not new, the application of the concept through the Multiple-Use Sustained-Yield Act of 1960, passed by the U.S. Congress, offered a much wider scope for public engagement with forest management and planning. After initially worrying about the potential negative impact of the Multiple-Use Sustained-Yield Act (it could lead to declines in timber production), USFS leaders embraced the policy as a means of assuaging public demands and fulfilling multiple objectives.

The Sierra Club resisted the passage of Multiple-Use Sustained-Yield Act of 1960 because its members worried that multiple use would not protect many of the most prized, un-logged forests from logging and road development. Through lobbying by preservationists, Congress passed the Wilderness Act of 1964, which legislated the creation of 9.1 million acres of wilderness, most of it located within USFS forests. The USFS (as well as the National Parks Service) opposed this act because it meant putting aside large swaths of productive USFS forestland in order to achieve a National Wilderness Preservation System. The legislation closed off large swaths of USFS from logging and allowed only limited research and recreation in the areas. The Wilderness Act of 1964 shifted the mandate of the USFS away from the original conservation model, which emphasized sustained-yield logging in almost all forests, toward a bifurcated model emphasizing timber production on certain lands and strict environmental protection on others.

In the 1960s, members of the Sierra Club began demanding the creation of a new national park to protect the last remaining old-growth redwood forests in California that were located on privately owned lands.[44] The battle over redwoods played a significant role in shaping negative attitudes toward the forest industry. It also opened up new ruptures between foresters and environmentalists who challenged the scientific and moral basis of scientific forest management.

Only 10 percent of redwoods were outside state lands on privately owned redwood forests, much of it devoted to commercial logging. Privately owned commercial forests employed foresters to manage redwoods under the American Tree Farms System. Commercial operators managed redwoods using "industrial forestry" operations to produce optimum yields while ensuring regeneration.[45] Some timber companies began noticing in the 1960s that "transient trespassers have been and continue to be a problem."[46] Visitors came to the privately owned forests because they offered "some of the best scenic and recreation areas in the region." Tensions between public visitors and private companies flared in the early 1960s as a result of the Arcada Redwood Company's decision to clear-fell 823 acres of redwood forest in sight of the Redwood Highway.[47] In the following years, almost every other timber company started to clear-fell redwood forests. The decision to use clear-felling paralleled the increased use of the practice in U.S. national forests.

When the Arcada Redwood Company started to clear-fell some of its forests, this set off a chain reaction of events. Sierra Club members, catalyzed by its energetic executive director David Brower, launched public attacks on private companies that logged redwood forests using clear-felling. Brower's more abrasive and aggressive approach to conservation separated him from his predecessors. Brower and his followers challenged foresters and timber companies directly. Rather than working quietly with American political executives, Brower appealed to the masses using language that rejected the technological and scientific supremacy of experts. In doing so,

Figure 3.5

Picturesque postcard showing giant sequoias in California

Credit: Photo courtesy of the Forest History Society, Durham, N.C.

he threw off almost a century of consensus conservation politics.

One of Brower's great insights was to use colorful advertising campaigns to shape public opinion about redwoods. Emanuel Fritz, professor of forestry at the University of California-Berkeley, believed that the Arcata Redwood Company's decision to begin clear-cutting its land gave environmentalist "proponents a much-needed opportunity for effective photographs to bolster its claim that the redwoods are fast disappearing."[48] Under Brower's leadership, the Sierra Club released a popular book, *The Last Redwoods* (1963), and produced a highly successful major ad published in the *New York Times*, *Los Angeles Times*, *Washington Post*, *San Francisco Chronicle*, and *Sacramento Bee*. In his autobiography, he mused, "Following the Redwood National Park and Grand Canyon ads, I had a hand in many others . . . They are written more carefully than the Bible, I will say, perhaps exaggerating to make the point."[49]

The conflict over redwoods inspired a new generation of environmentalists belonging to a range of revamped and new organizations—the Sierra Club (founded in 1892), Wilderness Society (founded in 1935), and EarthFirst! (founded in 1979)—that developed a more aggressive tactic to stop logging. Brower utilized publicity campaigns of "raw clearcuts" to shock the public, a goal he accomplished.[50] These organizations served as a vanguard, and publicly visible group, protesting forestry decisions throughout the country—especially in the American West.

Clear-felling touched a deep nerve in many Americans in the 1960s and 1970s. Criticism came from people of diverse backgrounds, ranging from radical environmentalists to conservative hunters. One of the most important fights over logging happened during the 1960s and 1970s as a result of clear-cutting in the Monongahela National Forest in West Virginia.[51] Hunters, hikers, and local business owners became alarmed when the U.S. Forest Service began clear-felling mature second-growth hardwood forests in 1964. USFS officers had not consulted with locals, and the cutting was done in an unsightly way. Hunters worried about how the cutting would affect populations of deer, turkey, and squirrel. Protests by local leaders, who had no affiliation with major environmental organizations, led the legislature of West Virginia to symbolically pass legislation in 1964 and again in 1967 and 1970 calling for an inquiry into clear-felling. When the USFS and federal government did not act on this, hunters from the West Virginia chapter of the Izaak Walton League of America successfully sued the USFS with a decision in 1975 calling for an end to clear-felling.

Even blue-collar workers and some professional foresters, supposedly traditional allies with extractive forestry, took up arms against clear-felling. The battle over harvesting in the Bitterroot National Forest was led by Guy Brandborg, the former U.S. Forest Service supervisor there from 1935 to 1955.[52] Brandborg feared that clear-cutting would negatively impact local communities in the long run by eroding the sustainability of its timber supply through over-extraction. In the Appalachians, there was strong resistance from local communities that had used national forests as a form of public commons. This tradition had deep roots within local communities who valued nearby national forests as places that they used for recreation and hunting.[53]

In response to public pressure, Congress passed the National Forest Management Act of 1976, which required public participation and interdisciplinary approaches when determining forest management plans. Yet the vague wording of the National Forest Management Act noted that it should be used "only where it is determined to be the optimum method" that aligned with general use. For this reason, conservation groups, ranging from the Sierra Club to the Izaak Walton League, did not support the bill.

Figure 3.6

Clear-felling Douglas fir in Pacific Northwest

Credit: Photo courtesy of the Forest History Society, Durham, N.C.

USFS extraction rates from national forests remained at historic highs from the 1960s until the late 1980s. The USFS was responding not only to its own internal scientific views, but also to lobbying by the timber industry and politicians, who saw timber sales from national forests within their electorate as a way to promote local jobs and boost political support. Foresters no doubt misjudged the public's response to clear-felling, but many members of the public misidentified foresters as the origin of the problem, when in fact they often followed dictates set by legislatures and lobbyists.

Clear-felling expanded in Australia from the late 1960s to the early 1970s. Australia's forest industry experienced an export boom in the 1970s. Japan's lack of domestic timber production and its growing economy required it to import huge amounts of timber, especially wood chips and fiber. The opening of a Japanese market led to a spike in domestic harvesting in state-controlled forests. Exports of wood chips shot up from less than 1 million cubic meters per year to 5 million by 1980, a fivefold increase in two decades.[54] The export boom led to the creation of pulp mills and export facilities along the coast in New South Wales, Tasmania, Victoria, and Western Australia. State foresters justified the use of clear-felling because it supposedly mimicked fire disturbances in native forests and was the most economical way of harvesting low- and high-grade timber. John Dargavel notes, "No one in Australia expected serious conflicts about exporting woodchips."[55]

Intellectual objections appeared soon after logging and exporting operations began. In their book *Fight for the Forests*, a seminal polemic work against clear-cutting in Australia, two Australian National University philosophers, Richard and Val Routley, carefully analyzed the Australian situation and drew attention to parallels with developments in America. The Routleys sought to philosophically undercut the basis of what they describe as "intensive forestry," which included but was not limited to clear-felling. They criticized foresters for trying to "evade conservation criticism mainly by retort to a set of pseudo-ecological arguments."[56] They particularly blasted clear-cutting, which they argued can "have a drastic and bad effect on forest ecology."[57] They paid particular attention to American developments, noting that "the Americans have been longer engaged on quite similar sets of issues . . . particularly intensive forestry practices, and their battle provides a fine preview of the local scenario."[58] Their influential text introduced the American problem to many Australian readers for the first time.[59]

Plans by the New Zealand Forest Service to clear-fell some of the South Island's last remaining native forests led to widespread public agitation against forestry. In 1971, the New Zealand government issued a white paper, "Utilization of South Island Beech Forests," in which the government proposed to cut 339,531 acres of beech forest on the South Island, primarily using clear-felling.[60] After being cut, the New Zealand Forestry Service would replant 50 percent the cutover land with exotic *Pinus radiata,* and 25 percent with *Eucalyptus*, leaving only 25 percent for native forest regeneration. Public and scientific criticism of the project led the government to abandon clear-felling in beach forests in 1975, but to push forward with its plan to log native forests in central North Island and the West Coast of the South Island. In response, environmental groups, scientists and general members of the public gathered together 341,160 signatures on the Maruia Declaration, which was handed to New Zealand's parliament in 1977. The principles of the Maruia Declaration called for an end to logging in "virgin forests," to stop burning after logging, and for the protection of native forests. One of the other principles called for the production of faster-growing exotic timber on non-forested lands to offset pressure on native forests.

Public discontent over forestry policies also occured in India, Thailand, Brazil, and Kenya during the 1970s and 1980s.[61] These protests focused less on clear-felling (which was an issue) and more on access to forests and the local benefits provided by forests. People often used nonviolent protest that sought to resist government indifference to the plight of people living in and near forests. The most famous example of resistance happened in the Himalayan districts of Uttarkhand and Uttar Pradesh in northern India, where peasants, particularly women, began to "hug" trees in the 1973 and 1974 to stop them from being harvested. The peasants protested against harvesting by commercial loggers, who won contracts from the Indian Forest Service and brought in their own skilled laborers. Local people gained little from forest harvesting, and the profits went to commercial firms and the Indian Forest Service. These protests played a particularly significant role in leading to the development of social and community forestry schemes in the 1980s to 2000s.

Change

The second half of the 1980s and early 1990s witnessed a turning point in the management of native forests in the United States, New Zealand, and Australia. Powerful popular, political, and scientific pressure led to the closing-down of many native forests, especially those that housed "old-growth" forest. New forest management practices, including "hands-off" policies for many forests, restructured the way foresters conceived of their roles. In all three cases, the decline of native timber harvests was offset by increased production from imports as well as private and government plantations, many that had been established between the 1930s and the 1960s. The success of timber plantation efforts nationally and globally helped make possible the transition away from native timber consumption. The trend of shutting down forests that occurred in these Western countries was mirrored in Thailand, China, South Africa, and Indonesia in the late 1980s to early 2000s

The legal mechanisms through which forests were set aside differed greatly in Australia, New Zealand, and the United States, though the timing and reasons for doing so were largely similar—pressure groups and perceived public opinion turned against logging in native forests, especially when logging was equated with clear-felling. In the United States and Australia, obscure legislative clauses were used to legally stop logging in federal and state forests. In Australia, the federal government used the "External Affairs" clause of its federal constitution to set aside forests in Tasmania and then Queensland. Stringent court orders based on federal legislation forced the USFS and private companies to stop logging on millions of hectares of forests in the Pacific Northwest of America. The decision by the Labour government of New Zealand to stop logging in state forests was more straightforward because there Parliament had authority over government forests.

In the early 1980s, political conflicts over resource development intensified in Australia, especially in the island state Tasmania, which housed some of Australia's largest old-growth forests in the southwestern part of the state. The conflict over logging in Tasmania pitted local residents against each other, as well as a pro-development state government versus the federal government, which sought to stop development in environmentally sensitive areas. Unlike the United States, where the federal government controls large swaths of land and has constitutional power to set laws nationally, Australian states have control over their resources, including forests and waterways. Throughout the early 1980s, both the Labor and Liberal parties in Tasmania supported the expansion of logging and the development of hydroelectric power in its western wilderness areas as a means to promote economic growth.

National public opinion turned against the Tasmanian state government's proposal to dam the Gordon River for hydroelectricity. In protest, the federal Coalition government applied to UNESCO in 1981 to list five of Tasmania's national parks as a single World Heritage site. Australia's parliament had in 1974 ratified UNESCO's World Heritage Convention (1972), which created rules for preserving sites of "outstanding universal value." Even if UNESCO's World Heritage Committee approved the site, which it was likely to do for political and scientific reasons, the federal government could not stop the Tasmanian state from building the dam because the constitution gave states control over forests and rivers.

The dam became a major federal political issue. In 1982, the World Heritage Committee approved the South West Tasmania World Heritage site but warned that it could be put on the "List of World Heritage in Danger" if Tasmania continued with the dam. In response, Tasmania's government pledged to go ahead with the project. Before the 1983 election, the opposition Labor Party pledged to stop the dam if it won. Though the incumbent federal Coalition government also wanted to stop construction, it felt the federal government lacked the constitutional powers to do that. Environmentalists supported Labor by running stunning photographs of the wilderness in newspapers before the election. After Labor won the federal election (without winning a single seat in Tasmania), Labor moved to stop Tasmania by drafting and passing the World Heritage Properties Conservation Act 1983, which forbid development in World Heritage areas, including the Western Wilderness, without Commonwealth ministerial approval. The state of Tasmania disregarded Labor's federal legislation by arguing that there was no provision in the Australian constitution for the federal government to intervene in state environmental affairs.

The case went to the High Court, which ruled in the 1983 decision *Commonwealth v. Tasmania* that the federal government had the right to interfere with states when it involved international agreements.[62] The landmark ruling provided a legal precedent that allowed the federal government to use UNESCO World Heritage listing to force states to comply with World Heritage rules on conservation. Because of Tasmania's continued efforts to log old-growth forests outside the parks, the Labor government expanded the World Heritage boundaries in 1989 to include forests not originally listed in 1982. This decision was greeted by much of the public as a great success, though Tasmanian state MP Bob Brown, the leading green politician of the time, derided the government for not going far enough.[63] The Labor government used the ruling to list threatened forests, including 900,000 hectares of the last remaining wet rain forests in northern Queensland (Wet Tropics of Queensland World Heritage Site, listed in 1988).[64]

New Zealand experienced an even more wide-sweeping change to its national forestry policies toward native forests from the mid-1980s to the mid-1990s. During the run-up to the federal election of 1984, the Labour government promised to stop logging in the native forests on the North Island as well as to review policies for managing indigenous and planted forests throughout New Zealand. After winning the election, recommendations by the government-appointed Commission for the Environment led to the closing down of the Forestry Service in 1987 and the establishment of two new departments, the Ministry for the Environment, which protected and managed native forests, and the Ministry of Forestry, which played a purely advisory role on forest extraction and policy. All state-controlled native forest on the North Island were deemed as protected under the Ministry for the Environment, and logging was only permitted on select forests on the west coast of the South Island. Remaining government-owned timber plantations were transferred to the New Zealand Forestry Corporation, created in 1987 as a for-profit corporate entity owned by the government. In 1989, the Corporation's assets, which included 550,000 hectares of plantations, began to be sold to various private interests and handed over to indigenous Maori. In 1993, a study by W. J. Hurdich pointed out that "multiple use as a management objective for public timber-production forests in New Zealand no longer applies."[65]

Tensions surrounding clear-felling and harvesting in native forests peaked in the western United States in the late 1980s and early 1990s during the spotted owl (*Strix occidentalis caurina*) controversy in the Pacific Northwest. The northern spotted owl is a nocturnal bird of prey that lives in old-growth forests in the Pacific Northwest. Little was known about the habits or population of the reclusive bird. USFS research in the early 1970s indicated that the population of northern spotted owl could decline with continued logging. The late 1980s saw elevated levels of harvesting in Oregon's federal forests, especially in old-growth Douglas fir forests. Ecological researchers became concerned that northern spotted owl lived in old-growth forests, the same forests that were prioritized for harvesting based on the theory that old-growth should be liquidated to make way for younger, more vigorous trees.

In the 1980s, environmental activists became disillusioned with major environmentalist groups, such as the Sierra Club, which were seen to be too close to industry and the USFS. Splinter groups, such as the "zero-cut" group," within the Sierra Club and more radical groups like EarthFirst! pursued new tactics to stop harvesting. These tactics including tree sittings (sitting in a tree so it cannot be cut), destroying logging equipment (called "Monkeywrenching" after the 1975 book *The Monkey Wrench Gang* by Edward Abby) and even tree spiking (i.e., putting a nail deep into a tree that will destroy a chainsaw and possibly hurt, or kill, the person holding it).

Grassroots environmental organizations turned to the federal courts to rule against the USFS and U.S. Fish and Wildlife Service over violations of the National Forest Management

Act, the National Environmental Policy Act, and, finally, the Endangered Species Act. One of the smallest environmental organizations played one of the biggest roles in lowering the harvest rate in Pacific Northwest forests. The tiny Green World environmental group petitioned the U.S. Fish and Wildlife Service in 1987 to have the northern spotted owl added to the endangered species list. After the initial request for a review was declined, the Sierra Club Legal Defense Fund (a separate entity from the Sierra Club) appealed the decision. In 1989, the District Court Judge William Dwyer, drawing on the ESA and the National Forest Management Act, ruled that the USFS had to design a viable plan to maintain the population of the northern spotted owl. Two years later Dwyer made another decision, stopping timber sales on lands that supported spotted owl. In making the 1991 decision, Dwyer wrote, "This is not the doing of scientists, foresters, rangers, and others at the working levels of these agencies. It reflects decisions made by higher authorities in the executive [i.e., presidential] branch of government."[66] Dwyer specifically referred to President George Bush's friendliness with the timber industry, which lobbied for higher extraction limits.

By the late 1980s, Dale Robertson, the twelfth forest chief of the USFS (1987–1993), believed that clear-felling was the primary issue turning the public against the U.S. Forest Service. Clear-felling was still supported by the timber industry as well as many senior forestry educators and old hands in forestry who came to believe that it was the most effective way to manage a range of species. In 1990, Robertson piloted a "New Perspectives" program that viewed forestry from a whole "ecosystem perspective."[67] Critics of forestry saw the program as a smoke-and-mirrors screen to hide business as usual, but it also featured ecological research focused on old-growth forests and unusual silvicultural regimes. Bush announced a new national forest policy during his visit to the United Nations Earth Summit in Rio de Janeiro in 1992. He declared that the federal government would phase out the use of clear-felling in national forests. Bush and his advisors made the hasty decision as a way to rebut criticism of the Republican government's environmental policy.[68]

U.S. federal forests witnessed a dramatic decrease in the harvest rate after 1992. USFS modeling predicted that in 1990 it would allow 11.1 billion board feet to be harvested from its forests, 10.8 billion board feet in 1995, and 12 billion in 2040.[69] In reality, reforms brought about during President Bill Clinton's first term in office cut the harvest rate dramatically to 3.1 billion board feet, a decrease of almost 75 percent from the late 1980s and a figure almost inconceivable even two years earlier. The most significant piece of this national policy was the Northwest Forest Plan, designed by Jack Ward Thomas, the thirteenth forest chief of the USFS. The Northwest Forest Plan was a multi-agency agreement to manage the 10 million hectares of forest in the Pacific Northwest within the range of the northern spotted owl.[70] In the aftermath of these changes, there was serious public discussion about whether the USFS should be abolished.[71]

Government decisions in China and Thailand to ban logging in degraded native forests paralleled changes in forest policy in the United States, New Zealand, and Australia. China and Thailand each faced a similar problem: decades of forest degradation through legal and illegal over-extraction was blamed for causing erosion and flooding. Unlike that of Australia or the United States, public criticism of government forest policy was focused less on retaining ecological purity of native forests and more on maintaining forest cover to provide a range of environmental services.

Thailand's government-controlled forests had been gradually cut as a result of the country's economic modernization during the 1950s and early 1980s.[72] In 1985, the Thai military government under the direction of General Prem Tinsulanonda instituted a logging ban that would see 15 percent of the

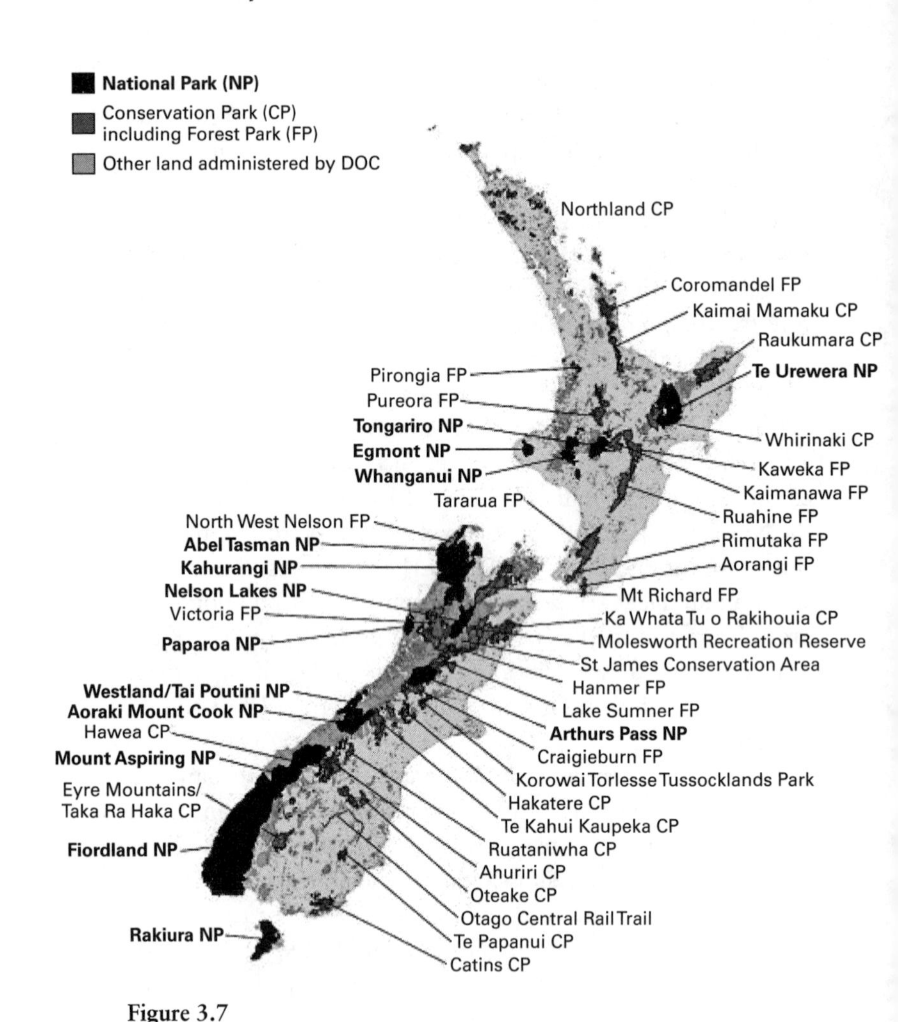

Figure 3.7

National parks and forest areas, New Zealand

Credit: Department of Conservation, New Zealand, http://www.doc.govt.nz/about-doc/role/maps-and-geospatial-services/.

country's remaining native forests go into protected areas. In 1988, southern Thailand suffered terrible floods. Critics blamed the flooding on the logging of steep slopes and the mountains.[73] As a response, the Thai government announced a wider logging ban in state-reserved forests in January 1989. With assistance from the United Nation's Food and Agricultural Organization, the government encouraged small agricultural holders as well as corporations to create eucalyptus plantations to partially make up for lost production of native forests.

Thailand's decision was followed ten years later by the Chinese government's decision to ban logging in large sections of its national forests and to focus efforts on increasing timber plantations. China's state-owned forests had undergone significant policy and ecological fluctuations from the late 1970s to the late 1990s as a result of the country's opening up to a market economy following the liberal reforms of Deng Xiaoping. China's government owned forestland, but its forest sector was inefficient and oversight over harvesting in forests led to high rates of illegal and legal harvesting. The government officially promoted tree planting as a way to stop desertification in its western hinterland and to generate timber through fast-growing exotic plantations.

A severe drought hit China in 1997, followed by heavy rains in 1998. The rains led to major flooding across China, especially along the Yangtze River. Public frustration led the government to quickly implement a National Forest Protection Program in 1998 that banned harvesting in degraded forests, on steep slopes, and near the headwaters of rivers. The policy sought to limit the harvest in native forests by a third, from 32 million cubic board feet meters to 12 million, and to place up to 90 million hectares of national forest into protected areas.[74]

On paper, China has largely succeeded in its goals, though it has only been able to relieve pressure on its own native forests by pursuing an aggressive sourcing of timber overseas—through legal and illegal means—that is causing extensive deforestation throughout Southeast Asia and Africa.[75] Like Japan, China does not have a large enough domestic timber supply to meet demand. China's rapacious appetite for tropical hardwoods is one of the greatest threats to tropical hardwoods. A 2014 report argues that in 2013, 90 percent of the timber harvested in Mozambique was cut illegally and exported to China.[76]

The End of Sustained Yield Forestry?

By the beginning of the twenty-first century, national forestry policies in countries throughout the world evidenced a clear shift toward producing timber in intensive plantations and lowering harvests in government-owned forests, either to preserve virgin forests or to regenerate degraded forest. There are, of course, exceptions to this rule. The Canadian province of British Columbia has continued—controversially—to utilize clear-felling to manage its extensive conifer forests.[77] The large volume of British Columbian softwood exports to the United States helps explain why the declining harvests in the Pacific Northwest since the early 1990s have had little impact on the supply of softwood timber in the United States.

Despite exceptions, it is clear which way the global trend is heading: we are moving toward a world that separates forest protection from timber production. This does not mean that countries with large native forests, such as Canada or Russia, will not utilize these forests. However, it is likely that developing countries with remaining native forests (such as Russia, Mozambique, and Indonesia) will deplete forest resources for export earnings, and developed countries (e.g., Canada and those in Scandinavia) will concentrate their efforts on harvesting in smaller areas more intensively while putting large forests into protected areas.

4

Toward a Twenty-First-Century Consensus: Problems and Possibilities

The Tasmanian state parliament passed the Tasmanian Forest Agreement in 2013 after decades of bitter fighting over how to manage the island's world-famous native forests. The agreement brokered by the state and federal Labor and Green Parties put aside approximately a half million hectares of native forest into protected status while offering secured rights to 137,000 cubic meters of native timber in addition to giving the timber industry and its workers over AU$300 million in compensation to help downsize and reorient the industry toward plantation-based production. It was hailed as a landmark agreement to solve the bitter "forest war" that polarized the state. Leaders of timber companies, timber worker unions, environmentalists, state foresters, and government officials all backed the agreement in the hope that it would provide a sustainable economic and ecological future for Tasmania. Yet this hope was short-lived. The agreement was repealed only a year later in 2014 by the incoming conservative Liberal government. Tasmania's forest future is now once again hotly contested.

Many environmentalists believe that the state's decision to reopen large swaths of native forest for logging is a last-ditch attempt to harvest many of Tasmania's remaining native forests, some that are centuries old, in order to prop up a declining timber industry that is being undercut by cheaper foreign competitors. They point out that it is cheaper for private Tasmanian timber companies to harvest from native state-owned forests than to grow trees on privately owned land because the state sells its timber to private contractors at a rate that is below the market rate for domestically grown plantation timber. Tasmania's government is now facing considerable pressure as a result of its decision to scrap the agreement. Critics of the current government policy argue that foreign and domestic investors will shy away from investing in the Tasmanian timber industry as a result of public opposition and questions about the future supply of native timber. Similar concerns played a role in forcing Tasmania's largest timber company, Gunns, to go into receivership in 2012.

The global and local forces shaping Tasmania's forest policy future are paralleled elsewhere in the world. Harvesting in native forests remains a contentious issue in developed and developing countries. The conflict over logging is often, but not always, fragmented along political lines, with conservative governments supporting an expansion of harvesting and left-leaning governments seeking to restrain and curtail harvesting. The future of timber production is uncertain in part because of this political division. At the same time, all but the largest timber producers in developed countries are finding it more difficult to compete against foreign rivals who are benefiting from advantageous environmental and economic conditions that allow them to produce cheaper timber for the global market. The economic revaluing of wood fiber has an impact on all aspects of forestry policy, from protection to production.

Forest Management at a Crossroads

Since the dawn of the twenty-first century, the globalization of timber production and processing has been one of the primary causes driving the disintegration of the conservation model and encouraging the forest management divergence, which is characterized by the proliferation of protected areas and intensive plantations. Market forces now disproportionately influence how and where the world's wood is produced. Demand for cheap wood fiber is driving the proliferation of exotic timber plantations throughout the developing tropics, where higher productivity and cheaper labor give growers an advantage in producing wood fiber over companies in temperate countries. The declining price and the greater abundance of cheap wood fiber are undermining the economic viability of small to medium-sized timber companies that rely on timber sales from state-controlled forests. The timber industry has seen a spate of national and multinational consolidations in order to achieve the international reach required to be competitive in the market. As a result of these economic changes, some economists and environmentalists now argue that governments should shut down harvesting in the majority of state-controlled native forests while relying on the private sector to produce timber.

While stakeholders from various perspectives agree that there is a problem with the existing policy framework, many worry that pursuing management plans that entirely separate forest production from protection will create new problems. Social and community forestry advocates point out that the creation of "untouched" protected areas continues to alienate local people from forests and to empower a new set of technical experts, such as ecologists and climate scientists, who determine what nature is and what it ought to be. Intensive plantations have been the cause of considerable social conflict in developing and developed countries. Some foresters warn that locking up forests is not a panacea because, without revenue to pay for proper management, they are prone to invasive species, climate change, fire, and other forces that can reshape them. Most protected areas are not properly funded and managed. Many people in rural areas support forest harvesting in state forests because it provides jobs. The idea of totally separating production from protection has little applicability in Europe, where large seminatural forests are oriented to production and protection. In countries such as China, with few remaining native forests, the bifurcation of production and protection is a response to an environmental catastrophe, rather than a well-planned policy to ensure sustainability.

I conclude by suggesting that stakeholders work together to advocate a new policy that engages interdisciplinary expertise and local communities to actively manage forests for resilience in the face of climate change, invasive species, and changing human visions of what nature is and ought to be. To do this requires that governments and markets take account of the externalities associated with offshoring timber production. The rise of free-trade liberalism, from a historical perspective, is recently new, and given the environmental problems it has caused it is possible that alternative trade arrangements, which are currently being directed by nongovernmental organizations (NGOs), may eventually shape the trade policies of countries. Putting a higher cost on timber and pricing in the externalities associated with the wood we purchase could help create incentives to recycle and allow small and midsized growers to profitably produce timber for the market in a more sustainable way. Harvesting in native forests that have already been managed should be allowed, but such harvesting should be directed by methods that increase or at a minimum sustain existing levels of biodiversity. Governments and universities should not allow the profession of forestry to fall into total disarray, because foresters are uniquely equipped to manage forests for protection and production. A key challenge of moving forward in the twenty-first century is to recognize past failures and to accommodate diverse needs without destroying all of the institutions and ideas that drove forest management for the past century.

Economic, Environmental, and Political Pressures

The combined effects of economic globalization and social protests against harvesting in native forests have reinforced a bifurcated policy focusing on timber production in intensive plantations and native forest preservation through protected areas. In some countries, notably China and New Zealand, these represented a conscious political decision. Yet many countries, such as the United States and Australia, are stuck between an older model that relied on harvesting in native forests at the same time that market and social pressures are encouraging a shift toward plantation production and the shutdown of native forests. A totally bifurcated system emphasizing plantations and protected areas is seen by some as a way to solve environmental and economic problems, although there are significant concerns that separating production from protection can lead to negative consequences.

The freer movement of wood products across national borders since the 1990s has been perhaps the single greatest factor shaping national forestry policies worldwide. The inflation-adjusted U.S. dollar value of the international timber trade grew from US$29 billion in 1961 to US$150 billion in 1999, and US$159.6 billion in 2009.[1] These figures do not reveal the true significance of this change because they do not account for the declining cost of timber as a result of increased efficiency in terms of growing, processing, and transporting timber. The rise in wood product imports in developed countries and growing consumption in developing countries is fueling the expansion of tropical plantations and even the illegal harvesting in biologically diverse forests in Asia, Africa, and South America.

Trade liberalization has occurred as part of a conscious decision by political elites to allow market forces to reshape economies and, by default, environments. Prior to the 1980s most governments tried to achieve timber independence by protecting forests and augmenting domestic production with plantations, if needed. But the ascendency of neoliberal thought, which called for the privatization of state assets and the tearing down of trade barriers, led political elites in certain countries during the 1980s to allow free-market mechanism to supply domestic timber needs. Neoliberal economic ideas inspired governments in Australia, New Zealand, and South Africa to begin privatizing state plantations in the mid- to late 1980s. Governments throughout the world took down tariff barriers on forest products as part of World Trade Organization (WTO) agreements. The U.S. government negotiated an elimination of tariffs on forest products (paper, and then later, furniture) with major trading regions (e.g., Canada, European Union, Hong Kong) at the 1993 Uruguay Round of the World Trade Organization.[2] There is still resistance to importing softwood lumber—as seen by the U.S.-Canadian controversy over softwood timber imports that ran from 1982 to 2006—but usually tariffs are invoked only if the volume of trade is massive and it directly threatens a large, organized industry.

Northern European and Canadian companies have sought to capitalize on trade liberalization by pursuing aggressive exports of paper, pulp, and softwood lumber to other developed countries. At the same time, Scandinavian timber companies have invested heavily in timber processing and plantations in developing countries, especially in South America. These investments have led to technology transfers that provide developing countries with access to cutting-edge knowledge developed in Finland, Norway, and Sweden, world leaders in wood processing technology. The search for overseas profits has led Scandinavian companies to partner with major South

American firms, which are blamed for dispossessing marginalized people of their land and encouraging the widespread planting of exotic species.[3] Scandinavia and Canada will likely remain leaders in selling softwood lumber and roundwood because of their vast forest cover, although there will be greater pressure to continue implementing more ecologically sensitive harvesting and regeneration programs.

The growth of foreign investment into developing countries, combined with growing imports into developed countries, has stoked a massive expansion in tropical timber plantations in developing countries, especially in South America and South and Southeast Asia (including China). Globally, plantations have grown at a rate of 3.6 million hectares per year during the 1990s and 4.9 million hectares per year during the 2000s.[4] Forest plantations constitute a sizeable ecological footprint—116 million hectares in Asia as of 2001, or 16 percent of its forest cover.[5] The increase in production from these plantations has seen a decline in the price of wood fiber used for a variety of industrial and consumer needs, especially goods purchased from flat-packing corporations, such as IKEA and Walmart.

Demand for cheaper wood fiber is a contributing cause (but it is by no means the leading cause) of deforestation in tropical developing countries both to provide timber and to make way for plantations. Peter Dauvergne and Jane Lister warn: "For natural forests, however, industrial plantations are a double-edged trend. In a lot of places, particularly in the global South, demand for commercial plantation land is driving deforestation."[6] There is concern that the inclusion of plantations as a legitimate form of "forest" in national United Nations Reducing Emissions from Deforestation and Forest Degradation (REDD+) framework may fuel greater deforestation of native forests in order to create more profitable intensive plantations.[7]

Even if successful, the creation of plantations is often associated with significant ecological risks. Plantations in tropical countries are composed primarily of exotic species grown in ecological and climatic conditions that differ greatly from the trees' indigenous conditions. In certain circumstances, introduced trees can become problematic invasive species that reshape entire ecosystems.[8] Exotic invasive trees, which were established as commercial or noncommercial plantations, are a significant ecological problem in many parts of the world. For instance, South Africa has a serious problem with invasive trees as a result of the legacy of its efforts to create plantations by planting exotic trees.

Tree planting in tropical regions is reshaping forest policy in developed countries. Declining wood prices are undermining the economic justification for managing native forests in Western countries. Governments in Canada, Australia, and the United States have been accused of selling native timber for a below-market rate to sustain industries that are not globally competitive. The most extreme response to these pressures is the call to "lock up" almost all native forests and to produce timber in intensive commercial plantations as part of a concerted policy. In Australia, the forest economist Judith Ajani has called for a "new forest policy that elevates plantation processing to the top forest industry goal and simultaneously ends logging in virtually all native forests."[9] This plan would allow native forests to be valued for protecting biodiversity, sequestering carbon, and safeguarding water. Timber would be produced as a commodity, not unlike minerals or agricultural products, which rely on the private sector for production.

There are some precedents for these reforms, but the extenuating circumstances of each situation raise questions about whether these examples can provide a general model for other countries. The exemplar of this bifurcated strategy is China, which is pursuing a plantations and protected areas policy on a grand scale—it currently has over 60 million hectares of plantation and 90 million hectares in protected forest (44 percent of its total forest cover).[10] Yet China is an autocratic centralized state that can dictate policy from the center without requiring democratic support. China also relies heavily on imports from other countries with less stringent environmental regulations. New Zealand also made sweeping reforms from the late 1980s to the late 1990s to protect all government-controlled native forests. There too, this was possible because the federal parliament, not smaller polities, controlled forests. Many countries are unable to make such sweeping decisions because of constitutional law (e.g., a weak federal government and strong states) or the difficulty involved in getting controversial legislation through the gridlocked politics of democratic assemblies.

It has been historically easier to create protected forest areas when native forests are degraded or when native forests are clearly unable to supply domestic timer needs. Thailand's and China's logging bans happened as a result of public attention directed at landslides and catastrophes that were linked to the overexploitation of government forests. Both governments decided to encourage plantations while using protected areas to regenerate native forests. In each instance, protected forest reserves are actually degraded forests that require regeneration, similar to forests in the American East in the early twentieth century. South Africa could justify transferring the management of the country's largest productive native forest, Knysna, to the National Parks System in 2005 because the forests produced (and continue to produce) only a tiny amount of the country's timber production. New Zealand's native forests were less productive than its vast plantations.

Yet there are reasons to be wary of any policy that locks away the majority of native forests in protected areas, something that many environmentalists would like to see happen. Ecologists recognize that a passive "hands-off" strategy of creating protected areas will not preserve ecosystems from human encroachment, invasive species, climate change, and habitat fragmentation.[11] The expansion of large protected areas globally has not been followed by an increase in funds to manage them.[12] Environmentalists have succeeded in setting aside land, but it is an open question whether or not they will succeed in preserving the values they want to maintain into the twenty-first century and beyond. A more active policy of management in parks and protected areas is necessary, but the expansion of protected areas has rarely been followed by a corresponding increase in funding required for active management. Unlike forestry, which often generated substantial revenues from harvesting, few protected areas receive enough tourism revenue or state funds to support active management. Foresters were also uniquely trained to both understand ecosystems and physically intervene and manage them.

Some serious questions remain about whether a separation of protection from production will work in tropical developing countries. Protected areas in developing countries are often protected only in name—in effect, as a "paper park." There are significant problems associated with corruption, illegal harvesting in protected forests, and often local disregard for national legislation in developing countries.[13] Attempts to stop deforestation through international investment and offset schemes, such as the UN's REDD+ program, have not succeeded as much as advocates had hoped, a significant blow to those who hope that internationally funded protected areas can save the rain forest.[14] One of the problems with trying to offset native forests with plantations is that deforestation, through legal or illegal means, is an attractive way of making money quickly without having to pay full rent value of land.[15] Nor have attempts to offset plantations with protected areas succeeded as well as hoped. A review of attempts to use timber plantations for conservation noted that "without concerted conservation efforts—for example, where weak governments are unable or unwilling to enforce conservation programs—plantations are less likely to reduce pressure on natural forests."[16]

Finding Balance in the Post-conservation Model World

I conclude this book by suggesting that the public and forest experts should be wary of embracing policies that entirely decouple timber production from forest protection. These policies reinforce the extremes of forest management, rely too much on free-market forces, and encourage the belief that forests have not historically been modified by human action in the past. Moreover, it decouples the creation of revenue from protection, leaving large swaths of forests without adequate management in the face of climate change, invasive species, continued ecological fragmentation, and increased pressure on the world's wood resources. Decoupling production from protection will continue to devalue the cost of native forest timber, forcing governments to either harvest using controversial techniques, such as clear-felling and wood-chipping, or to shut down harvesting altogether in native forests. Management plans need to be consultative, drawing together various stakeholders, and achieving multiple outcomes that are oriented to benefiting local, national, and global communities.

Unlike the conservation model, which was part of a considered policy decision, the forest management divergence has happened largely through a series of uncoordinated processes. Few foresters, environmentalists, or members of the timber industry would necessarily have chosen this path. Closing down native forests in many parts of the world has alienated the timber industry from native forests, which have historically provided much of their supply. The profession of forestry is in decline throughout much of the world as a result of social conflicts over harvesting in native forests. Environmentalists are concerned about the longevity of protected areas and the fate of the remaining 85 percent of the world's forests not located within a protected area. Local communities still often feel estranged from forests. No single interest group feels that it has "won," and there is a general sense of frustration on all sides.

To move forward all parties—industry, environmentalists, foresters, and local communities—must compromise to achieve a workable long-term policy. There is a large volume of research now on best practices that can be incorporated into an integrated framework. Governments need to spearhead policy discussions because they alone have the regulatory ability to create new legislative rules and to subsidize and tax new policies. The attempt by the Tasmanian Labor and Green coalition to create a forestry agreement in Tasmania, mentioned earlier, provides one such precedent. It is unlikely that there will be any one model of consensus given the varying political and environmental conditions throughout the world, but these processes should follow best practices of deliberative democracy.[17] Countries that lack democratic structures need to engage deliberatively with communities to ensure that any policies can survive regime change or decentralization in the future.

There is a greater need to decentralize decision making to hand over some management power to local communities, especially in developing countries. Though the evidence on community forestry efforts is checkered, many have pursued more ecologically sustainable policies than have their state forestry services.[18] These schemes are ideally suited to programs focused on ecological outcomes or the production of limited volumes of wood for local communities, especially those in timber-deficient regions, such as Nepal, one region where such forestry efforts have had some success. Mexico has a particularly rich and long history of community forestry efforts that have continued since before the Revolution in 1910 despite state appropriation of forest management decision making from the 1940s to the 1970s.[19] Community forestry efforts often work best in scenarios where forests are threatened by local deforestation or where local residents require forests for their livelihood. There are possibilities for matching community-managed forests with larger industrial needs.[20] It may be possible to link together community forestry efforts through forest certification schemes that give added value to schemes with favorable social and ecological conditions.

The economics of national forest management policies must be adjusted to reintegrate production and protection. The separation of plantations from protected areas often benefits the private sector at the expense of the public, who must carry the bill for protection without revenue from timber production. Protected areas require more funding for active management to stop the spread of invasive species and to intervene if climate change leads to species migration. Historically, harvesting has provided revenue for other forms of ecological management.

There are examples, such as the Knysna forest in South Africa, where funding from plantation-grown timber supported the protection and limited harvesting of the biologically diverse subtropical native forests. Yet in 2005 the South African government gave the Knysna forest to an already overextended National Parks system to be protected (the plan still allows for minimal harvesting of timber). For almost eighty years Knysna was managed using ecologically sensitive methods that allowed for minimal extraction and maximum environmental protection. This program was underpinned financially by the success of plantations surrounding the forest. There are questions about whether the cash-strapped National Parks has the resources to protect the forest from illegal harvesting of nonwood products and invasive species. Public pressure to close down logging threatens to destroy the region's small but culturally important wood processing and furniture industry.

Free trade is one of the elephants in the room that must be acknowledged. So long as current free-trade policies continue, little hope exists for a slowdown in the shift of wood production from temperate north to tropical south. Nongovernmental organizations have stepped in to regulate the market in lieu of stronger government regulation, which is directed primarily toward putting an end to illegal logging, the export of unfinished logs, and "dumping" large volumes of cheap softwood. There has been a proliferation of forest certification schemes that inform buyers that the timber meets basic sustainability standards. The most influential certification scheme is the international Forest Stewardship Council (FSC), but there are others with more stringent environmental certification. Forest certification is a means to help consumers buy wood that meets certain environmental standards that conform to the principle of sustainable forest management (SFM), a concept developed in 1992.[21] Yet NGOs and consumers can only do so much on their own. Governments need to intervene by creating policies that put more stringent requirements on timber imports. This is unlikely given support for free trade, but there is the possibility that in the coming decades environmental and social pressures may make this more desirable.

There is a need to maintain the existing professional competence of foresters, who have unique skills required to manage large forests. We may find ourselves in more serious trouble if we allow the profession of forestry to decline any further. Major forestry programs have closed or have merged with other disciplines and no longer teach key management practices relating to the harvesting of trees in native forests. Professional foresters have been caught out in the contradictory pulls of current policy. A policy submission to the Australian government noted: "There is growing demand for forests to be managed to provide a wider range of ecosystem services and social values, but our management capacity has diminished . . . our professional capacity in forest sector research, and in managing forests for multiple values, services and products, are declining."[22] Yet it should be clear that foresters will never regain the central role they once played in the older conservation model.

It may be desirable to move toward a "middle path," one that involves highly regulated selective harvesting in native forests, even many of those protected for ecological purposes in order to guard forests against climate change, invasive species, and ecological fragmentation. There is a wealth of ecological and silvicultural knowledge, including mixed wood systems, that can be economical and ecologically sustainable. Foresters, of course, are the strongest advocate of this view, but it's a view that has been echoed by scholars in many fields. This policy may be particularly effective in certain developing nations, where forests are felled illegally to make way for agricultural land.[23] Active forest management can protect against the buildup of fuel loads that can create intense, ecologically destructive fires. Active management can also fight against invasive species. If timber commands higher prices, selective management schemes become financially viable.

There are some hopeful signs that the profession of forestry has reinvented itself to fit the wider needs of society. The development of new scientific and people-centric perspectives has yielded positive results. Numerous examples of successful selective management programs that have been developed for a diverse range of ecological and social systems have emerged. So too have various schemes to decentralize forests *with* the help of foresters shown great promise. These developments all offer possible solutions to the problems related to the establishment of plantations, the management of protected areas, and harvesting in native forests.

There will no doubt be future assessments about the history of forestry as new problems threaten the world's forests. It is possible that we may once again find some value in the conservation model. Considered in terms of community participation and access, the conservation model was often less exclusive than strictly protected areas and intensive plantations, which seek a strict separation of people and forests. Government forestry programs usually mandated some type of multiple-use framework that offered some form of public participation. Undoubtedly incarnations of the conservation model, especially in colonial countries, created significant social problems arising from the public's inability to challenge and shape management decisions. One of the biggest problems with the conservation model was that professional foresters too often *determined* policies without public input rather than *collaborating* with communities and seeing government forests as a type of public commons.

No one-size-fits-all model exists for determining how a country, region, or landowner manages a forest. There is no doubt given population growth and pressure on the earth's systems that we will require greater outputs from plantations as well as better protection of native forests. But we should be careful to monitor the forest management divergence to make sure it does not cause problems equal to or greater than the conservation model. If the divergence between production and protection becomes too extreme, we may face a bleak future where the proliferation of protected areas is not matched by resources to maintain them and the offshoring of timber production to developing countries encourages deforestation and the expansion of intensive exotic timber plantations. Yet there are also reasons to be optimistic. Planned carefully, the dual trend of protected areas and plantations offers exciting potential to produce timber while protecting the world's ecosystems.

In order to achieve the best possible future outcome, it is critical that policy makers, scientists, environmentalists, and all interested people understand that the world is undergoing a forest management divergence that is reshaping timber production and forest protection. Only once we recognize this historical process can we then begin to resolve the tensions and utilize the possible synergies that are inherent in the two models, plantations and protected areas, which have emerged out of an older conservation model that is in decline globally.

Notes

Series Foreword

1. Gary Snyder, "Little Songs for Gaia," in *No Nature: New and Selected Poems* (New York: Pantheon, 1992), 287. Originally published in book form in Gary Snyder, *Axe Handles* (Berkeley: North Point Press, 1983).

2. http://www.ghostforest.org/.

3. Richard P. Tucker, *Insatiable Appetite: The United States and the Ecological Degradation of the Tropical World* (Lanham, MA: Rowman & Littlefield, 2007), 185.

4. I am "stealing" the conspiracy theory line from Paul Hirt, *A Conspiracy of Optimism: Management of the National Forests since World War Two* (Lincoln: University of Nebraska Press, 1994).

5. Diana K. Davis, *Resurrecting the Granary of Rome: Environmental History and French Colonial Expansion in North Africa* (Athens: Ohio University Press, 2007).

6. Paul Warde, "Fear of Wood Shortage and the Reality of the Woodland in Europe, c. 1450–1850," *The History Workshop Journal* 62, no. 1 (2006): 28–57.

7. Richard H. Grove, *Green Imperialism: Colonial Expansion, Tropical Island Edens, and the Origins of Environmentalism, 1600–1860* (New York: Cambridge University Press, 1995).

8. For more on sacrifice in this context, see Michael Maniates and John M. Meyer, eds., *The Environmental Politics of Sacrifice* (Cambridge, MA: MIT Press, 2010).

9. Madhav Gadgil and Ramachandra Guha, "State Forestry and Social Conflict in British India," *Past and Present* 123 (1989), 144.

10. Robert Wuthnow, *Be Very Afraid: The Cultural Response to Terror, Pandemics, Environmental Devastation, Nuclear Annihilation, and Other Threats* (New York: Oxford University Press, 2010), 1.

Introduction

1. Jerry Franklin, "Challenges to Temperate Forest Stewardship—Focusing on the Future," in *Toward Forest Sustainability*, edited by David Lindenmayer and Jerry Franklin (Collingwood: CSIRO Publishing, 2003), 1.

2. Ibid., 2.

3. Fredrik Albritton-Jonsson, *Enlightenment's Frontier: The Scottish Highlands and the Origins of Environmentalism* (New Haven: Yale University Press, 2013); Gregory Barton, *Empire Forestry and the Origins of Environmentalism* (Cambridge: Cambridge University Press, 2002).

4. A. G. Hopkins, ed., *Globalization in World History* (London: Pimlico, 2002).

5. James Scott, *Seeing like a State: How Certain Schemes to Improve the Human Condition Have Failed* (New Haven: Yale University Press, 1999), 19–20.

6. Ibid.

7. Ravi Rajan, *Modernizing Nature: Forestry and Imperial Eco-development 1800–1950* (Oxford: Oxford University Press 2006).

8. K. Sivaramakrishnan, *Modern Forests: Statemaking and Environmental Change in Eastern India* (Palo Alto: Stanford University Press, 1999), 243.

9. Ibid., 16.

10. David N. Livingstone, *Putting Science in its Place: Geographies of Scientific Knowledge* (Chicago: University of Chicago Press, 2003). A notable attempt to pursue this is James Beattie, *Empire and Environmental Anxiety: Health, Science, Art and Conservation in South Asia and Australasia, 1800–1920* (Basingstoke: Palgrave Macmillan, 2011).

11. See Charles Mann, *1491: New Revelations of the Americas before Columbus* (New York: Knopf, 2005).

1 The Conservation Model

1. A. G. Hopkins, ed., *Globalization in World History* (London: Pimlico, 2002).

2. One exception to this general rule is the work of Joachim Radkau. See Joachim Radkau, "Wood and Forestry in German History: In Quest of an Environmental Approach," *Environment and History* 2, no. 1 (1996): 63–76, 70.

3. Derek Wall, *The Commons in History: Culture, Conflict and Ecology* (Cambridge, MA: MIT Press, 2014).

4. See Peter Vandergeest, "Mapping Nature: Territorialization of Forest Rights in Thailand," *Society and Natural Resources: An International Journal* 9, no. 2 (1996): 159–175.

5. E. A. Wrigley, *Continuity, Chance and Change: The Character of the Industrial Revolution in England* (Cambridge: Cambridge University Press, 1990), 12. Fredrik Albritton-Jonsson, "Rival Ecologies of Global Commerce: Adam Smith and the Natural Historians," *American Historical Review* 115, no. 5 (2010): 1342–1363.

6. Albritton-Jonsson, "Rival Ecologies of Global Commerce."

7. Cited in Char Miller, *Seeking the Greatest Good: The Conservation Legacy of Gifford Pinchot* (Pittsburgh: University of Pittsburgh Press, 2013), 46.

8. Karl Appuhn, *A Forest on the Sea: Environmental Expertise in Renaissance Venice* (Baltimore: Johns Hopkins University Press, 2009), 95.

9. Paul Warde, "Fear of Wood Shortage and the Reality of the Woodland in Europe, ca.1450–1850," *The History Workshop Journal* 62, no. 1 (2006): 28–57, 42.

10. Thomas B. Van Hoof et al., "Forest Re-growth on Medieval Farmland after the Black Death Pandemic—Implications for Atmospheric CO2 Levels," *Palaeogeography, Palaeoclimatology, Palaeoecology* 237, nos. 2–4 (2006): 396–409.

11. Warde, "Fear of Wood Shortage."

12. Tamara Whited, *Forest and Peasant Politics in Modern France* (New Haven: Yale University Press, 2000); Richard Holzl, "Historicizing Sustainability: German Scientific Forestry in the Eighteenth and Nineteenth Centuries," *Science as Culture* 19, no. 4 (2010): 431–460.

13. Kieko Matteson.

14. See Robert Brenner, *Property and Progress: The Historical Origins and Social Foundations of Self-Sustaining Growth* (London: Verso, 2009).

15. Charles Watkins, *Trees, Woods and Forests: A Social and Cultural History* (London: Reaktion, 2014), chapter 7.

16. Jeremy L. Caradonna, *Sustainability: A History* (New York: Oxford University Press, 2014), 35–38. A Google books search of Carlowitz and sustainability turns up over one thousand diverse results.

17. Henry Lowood, "Patriotism, Profit, and Scientific Societies, 1760–1815" (PhD, University of California at Berkeley, 1987); Henry Lowood, "The Calculating Forester: Quantification, Cameral Science, and the Emergence of Scientific Forestry Management in Germany," in *The Quantifying Spirit in the Eighteenth Century*, ed. Tore Frängsmyr, J. L. Heilbron, and Robin E. Rider (Berkeley: University of California Press, 1990), 315–342.

18. Scott, *Seeing like a State*, chapter 1.

19. See John Dargavel and Elisabeth Johann, *Science and Hope: A Forest History* (Cambridge: White Horse Press, 2013).

20. For a foresters' view, see ibid. See discussion in the introduction and later in this chapter.

21. Scott, *Seeing like a State*, 19.

22. See, for instance, Bernard Fernow, *A Brief History of Forestry in Europe, the United States, and Other Countries* (Toronto: University of Toronto Press, 1907).

23. See, for instance, Donald Worster, *The Wealth of Nature: Environmental History and the Ecological Imagination* (New York: Oxford University Press, 1993), 144.

24. Paul Warde, "The Invention of Sustainability," *Modern Intellectual History* 8, no. 1 (2011): 153–170.

25. Sidney Mintz, *Sweetness and Power: The Place of Sugar in Modern History* (New York: Penguin Books, 1985), xxiv.

26. Conrad Totman, *The Green Archipelago: Forestry in Preindustrial Japan* (Los Angeles and Berkeley, University of California Press, 1989).

27. Jocahim Radkau, "Germany as a Focus of European 'Particularities' in Environmental History," in *Germany's Nature: Cultural Landscapes and Environmental History*, ed. Thomas Lekan and Thomas Zeller (Piscataway: Rutgers University Press, 2005), 17–32, 21.

28. Warde, "Fear of Wood Shortage," 49.

29. Scott, *Seeing like a State*, 19–20.

30. Janna Puumalainen, *Structural, Compositional, and Functional Aspects of Forest Biodiversity* (Geneva: United Nations, 2001), 23.

31. Julian Evans, "Introduction," in *Planted Forests: Uses, Impacts and Sustainability*, ed. Julian Evans (Oxford: CABI, 2009), 5–22, 8. See also chapter 2.

32. Heinrich Cotta, "Cotta's Preface," *Forest History Today* (2000): 27–28.

33. For instance, see William Schlich, *Schlich's Manuel of Forestry: Volume Two, Sylviculture*, 3rd ed. (London: Bradberry, Agnew, and Co., 1904), 72.

34. Michael Williams, *Deforesting the Earth: From Pre-history to Global Crisis* (University of Chicago Press 2003), 417.

35. For the origin of the term, see A. S. Mather, "The Forest Transition," *Area* 24, no. 4 (1992): 367–379.

36. Warde, "Fear of Wood Shortage," 51.

37. George Perkins Marsh, *Man and Nature: Or, Physical Geography as Modified by Human Action* (Cambridge, MA: Belknap Press of Harvard University Press, repr., 1965), 48.

38. See discussion by Diana Davis, *Resurrecting the Granary of Rome: Environmental History and French Colonial* (Athens: Ohio University Press, 2006), 75–78; Gregory Barton, *Empire Forestry and the Origins of Environmentalism* (Cambridge: Cambridge University Press, 2002).

39. Davis, *Resurrecting the Granary of Rome*, 77.

40. I use the date of the creation of forestry services or the passing of key forestry legislation. Many countries had no single date when forestry began. The point here is to illuminate the broader trend rather than to emphasize a specific date.

41. Thomas Jefferson, *Notes on the State of Virginia*, in *The Works of Thomas Jefferson Volume 3* (New York and London, G. P. Putnam's Sons, 1904), 481.

42. See James Flemming, *Historical Perspectives on Climate Change* (New York: Oxford University Press, 1998), chapters 1 and 2.

43. Richard Grove, *Green Imperialism: Colonial Expansion, Tropical Island Edens and the Origins of Environmentalism, 1600–1860* (Cambridge: Cambridge University Press, 1995).

44. Ibid., 198.

45. For the examples in this paragraph, see Gregory Cushman, "Humboldtian Science, Creole Meteorology, and the Discovery of Human-Caused Climate Change in South America," *Osiris* 26, no. 1 (2011): 19–44.

46. Today it is assumed that the lake's levels fluctuated due to normal variations in rainfall.

47. Jean-Baptiste Rougier de La Bergerie, *Mémoire et observations sur les abus de défrichements et la destruction des bois et forêts* (Auxerra, 1800).

48. Vazken Andréassian, "Waters and Forests: From Historical Controversy to Scientific Debate," *Journal of Hydrology* 291, no. 1 (2004): 1–27.

49. See "Biblical Botany," in *The Theological Eclectic*, ed. George E. Day (Cincinnati: William Scott, 1864), 94.

50. Marsh, *Man and Nature*, vi.

51. James Bellich, *Replenishing the Earth: The Settler Revolution and the Rise of the Anglo-World, 1783–1939* (Oxford: Oxford University Press, 2009).

52. Keith Hancock, *Australia* (London: Ernest Benn, 1930), 33.

53. See Warren Dean, *With Broadax and Firebrand: The Destruction of the Brazilian Atlantic Forest* (Berkeley: University of California Press, 1995).

54. Marsh, *Man and Nature*, 250.

55. Ian Tyrell, *Crisis of the Wasteful Nation: Empire and Conservation in Theodore Roosevelt's America* (Chicago: University of Chicago Press, 2015).

56. See Ezra D. Rashkow, "Idealizing Inhabited Wilderness: A Revision to the History of Indigenous Peoples in National Parks," *History Compass* 12, no. 10 (2014): 818–832.

57. Bernhard Gissibl, Sabine Höhler and Patrick Kupper, eds., *Civilizing Nature: National Parks in Global Historical Perspective* (New York: Berghan Books, 2012).

58. Sandra Chaney, *Nature of the Miracle Years: Conservation in West Germany, 1945–75* (New York: Berghahn Books, 2008), 215.

59. Heinrich von Salisch, *Forest Aesthetics* (Durham, NC: Forest History Society, 2008), 149–150.

60. Tyrell, *Crisis of the Wasteful Nation*, 146.

61. John Muir, "The National Parks and Forest Reservations," *Sierra Club Bulletin* (1896): 268–284, 284.

62. Scott, *Seeing like a State*, 19.

63. Ravi Rajan, "Imperial Environmentalism or Environmental Imperialism? European Forestry, Colonial Foresters, and the Agenda of Forest Management in British India, 1800–1900," *Nature and the Orient* (New York: Oxford University Press, 1998), 3. Cited in William Beinart and Lotte Hughes, *Environment and Empire* (Oxford: Oxford University Press, 2007), 126.

64. Madhav Gadgil and Ramachandra Guha, "State Forestry and Social Conflict in British India," *Past and Present* 123 (1989): 144.

65. Ravi Rajan, *Modernizing Nature: Forestry and Imperial Eco-development 1800–1950* (Oxford: Oxford University Press 2006), 84.

66. Ibid., 85.

67. Shepard Krech, *The Ecological Indian: Myth and Imagination* (New York: W. W. Norton, 2000). See Beinart and Hughes, *Environment and Empire*, 124–129.

68. For the most up-to-date review, see Christopher E. Doughty, "Preindustrial Human Impacts on Global and Regional Environment," *Annual Review of Environment and Resources* 38 (2013): 503–527.

69. Ibid.

70. Fernow, *A Brief History of Forestry in Europe*, 400.

71. Edward Stebbing, *The Forests of India Volume II* (Edinburgh: John Lane, 1923), 564–565.

72. Ibid.

73. Thaddeus Raymond Sunseri, *Wielding the Ax: State Forestry and Social Conflict in Tanzania, 1820–2000* (Athens: Ohio University Press, 2009), 79.

74. See Brett M. Bennett, "Naturalising Australian Trees in South Africa: Climate, Exotics, and Experimentation," *Journal of Southern African Studies* 37 (2011): 265–80; James Beattie, *Empire and Environmental Anxiety: Health, Science, Art and Conservation in South Asia and Australasia, 1800–1920* (Basingstoke: Palgrave Macmillan, 2011); Peter Vandergeest and Nancy Peluso, "Empires of Forestry: Professional Forestry and State Power in Southeast Asia, Part 2," *Environment and* History 12, no. 4 (2006): 359–393, 364.

75. Ian Tyrell, *Crisis of the Wasteful Nation: Empire and Conservation in Theodore Roosevelt's America* (Chicago: University of Chicago Press, 2015), 34.

76. Barton, *Empire Forestry*, 140.

77. Gifford Pinhchot, *Breaking New Ground* (New York: Harcourt Brace Jovanovich, 1947), 17–18. Cited in Barton, *Empire Forestry*, 140.

78. For Lane Poole, see John Dargavel, *The Zealous Conservator: A Life of Charles Lane Poole* (Perth: University of Western Australia Press, 2008). For Vasselot and Hutching, see Bennett, "Naturalising Australian Trees in South Africa."

79. Lawrence S. Earley, *Looking for Longleaf: The Fall and Rise of an American Forest* (Chapel Hill: University of North Carolina Press, 2004).

80. Peter Dauvergene and Jane Lister, *Timber* (Cambridge: Polity Press, 2011), 61.

81. Tom Griffiths, *Forests of Ash: An Environmental History* (Melbourne: Cambridge University Press, 2001), 37.

82. Barton, *Empire Forestry*, 58–59, 144, 159–162. In the United States, a landmark piece of legislation was the Multiple-Use Sustained-Yield Act of 1960.

83. See the debate over land tenure in South India before the implementation of the Madras Forest Act of 1882. Madhav Gadgil and Ramachandra Guha, *This Fissured Land: An Ecological History of India* (Berkeley: University of California Press, 1993), 124–132.

84. There is a large literature on this subject. For two excellent studies, see Ramachandra Guha, *The Unquiet Woods: Ecological Change and Peasant Resistance in the Himalaya*, 2nd ed. (Oxford: Oxford University Press, 2000), 48–131; Jacob Tropp, *Natures of Colonial*

Change: Environmental Relations in the Making of the Transkei (Athens: Ohio University Press, 2006), 31–123.

85. Berhard Fernow, *Economics of Forestry: A Reference Book for Students of Political Economy and Professional and Lay Students of Forestry*, 2nd ed (New York: Thomas Y. Crowell, 1902), 186.

86. Stephen Pyne, *World Fire: The Culture of Fire on Earth* (New York: Henry Holt, 1995), 29–170; Davis, *Resurrecting the Granaries of Rome*, 65. This trope is littered throughout history, forestry, and ecology publications.

87. Harold K. Steen, *The U.S. Forest Service: A History* (Durham, NC: Forest History Society, 2004), xiv.

88. Dargavel, *The Zealous Conservator*, 63–77.

89. Ibid.

90. Luc Bouthillier, "Quebec: Consolidation and the Movement toward Sustainability," in *Canadian Forest Policy: Adapting to Change*, ed. Michael Howlett (Toronto: University of Toronto Press, 2001): 237–278, 248.

91. The classic study is Herbert Kaufman, *The Forest Ranger: A Study in Administrative Behavior* (Baltimore: Johns Hopkins University Press, 1960).

92. Dietrich Brandis, *Indian Forestry* (Woking: Oriental University Institute, 1897), 35.

93. I. J. Craib, *The Wattle Industry in South Africa, British Empire Forestry Conference, South Africa, 1935* (Pretoria: Government Printer, 1935).

94. Rajan, *Modernizing Nature*, 171–179.

95. Thomas W. Webber, *The Forests of Upper India and Their Inhabitants* (London, 1902), 28–29.

96. Michelle Williams, *Deforesting the Earth: From Prehistoric to Global Crisis, An Abridgement* (University of Chicago Press, 2010), 372.

2 Plantations

1. Suhkum Thirawat, *The Eucalypts for Tropical Climates: Based on Experiences Gained from the FAO Eucalyptus Study Tour in Australia 1952* (Bangkok: FAO, 1952), 1.

2. Madhav Gadil and Ramachandra Guha, "Ecological Conflicts and the Environmental Movement in India," *Development and Change* 24 (1994): 101–136, 105.

3. Chris Baker and Pasuk Phongpaichit, *A History of Thailand*, 2nd ed. (Cambridge: Cambridge University Press, 2009), 217.

4. FAO, "Global Forest Resource Assessment," *FAO Forestry Paper 163* (Rome: FAO, 2010), 90; FAO, *Planted Forests in Sustainable Forest Management: A Statement of Principles* (Rome: FAO, 2010), section 3.

5. The 2010 Forest Resource Assessment [FRA] did not include this statistic. See the 2000 report. FAO, "Global Forest Ressources Assessment 2000," *FAO Forestry Paper 139* (Rome: FAO, 2001).

6. Rubin Shmulsky and P. David Jones: *Forest Products and Wood Science: An Introduction*, 6th ed. (Chichester: John Wiley and Sons, 2011), 451; Roger Sands, *Forestry in a Global Context* (Oxford: CABI Publishing 2005), 91–95.

7. Chris Brown and James Ball, "World View of Plantation Grown Wood," *Forestry Department Food and Agriculture Organization of the United Nations Forest Plantations Thematic Papers* (Rome: FAO, 2001), 6.

8. Mordechai Kislev, Anat Hartmann, and Ofer Bar-Yosef, "Early Domesticated Fig in the Jordan Valley," *Science* 312, no. 5778 (2006): 1372–1374.

9. John Richards, *The Unending Frontier: An Environmental History of the Early Modern World* (Berkeley and Los Angeles: University of California Press, 2005), 129.

10. Conrad Totman, *Japan: An Environmental History* (London: I. B. Taurus, 2014), 178–179.

11. Conrad Totman, *The Green Archipelago: Forestry in Preindustrial Japan* (Berkeley: University of California Press, 1989), 6.

12. Heinz Ellenberg, *Vegetation Ecology of Central Europe* (Cambridge: Cambridge University Press, 1988), 526.

13. See A. T. Grove and O. Rackham, *The Nature of Mediterranean Europe: An Ecological History* (New Haven: Yale University Press, 2001).

14. John Croumbie Brown, *Pine Plantations on the Sand-Wastes of France* (Edinburgh: Oliver and Boyd, 1878), vii.

15. Diana Davis, *Resurrecting the Granary of Rome: Environmental History and French Colonial Expansion in North Africa* (Athens: Ohio University Press, 2007), 102–108.

16. John E. Brown, *A Practical Treatise on Tree Culture in South Australia* (Adelaide: E. Spiller, Government Printer, 1881), 4–9.

17. R. N. Parker, *Eucalyptus in the Plains of North West India* (Calcutta: Government of India Printer, 1925).

18. David Hutchins, "Extra-Tropical Forestry: Being Notes on Timber and Other Trees Cultivated in South Africa and in the Extra-Tropical Forests of Other Countries," *Agricultural Journal of the Cape of Good Hope* 26, no. 1 (1905): 521.

19. Shaul Cohen, *Planting Nature: Trees and the Manipulation of Environmental Stewardship in America* (Berkeley and Los Angeles: University of California Press, 2004), chapter 3.

20. Jodi Frawley, "Campaigning for Street Trees, Sydney Botanic Gardens, 1890s–1920s," *Environment and History* 13 (2009): 303–322; Libby Robin, *How a Continent Created a Nation* (Sydney: UNSW Press, 2007), 13–23.

21. For the statistics for Brazil and South Africa, see Julian Evans, *Plantation Forestry in the Tropics*, 2nd ed. (Oxford: Oxford University Press, 1992), 26–27; for New Zealand, see J. Powell, "The Empire Meets New Deal: Interwar Encounters in Conservation and Regional Planning," *Geographical Research* 43, no. 4 (2005): 337–360, 345; for Britain, see E. Richards, *British Forestry in the Twentieth Century: Policy and Achievement* (Leiden: Brill, 2003), xii.

22. As early as the 1920s, private growers produced 70 percent of South Africa's plantation timber supply of wattle and eucalyptus. W. Duncan Reekie, "The Wood from the Trees: *Ex Libri ad Historiam Pertinentes Cognoscere*," *South African Journal of Economic History* 19, nos. 1–2 (2004): 67–99, 73–74.

23. Gifford Pinchot, *The Fight for Conservation* (New York: Double Day, 1910), 15.

24. 1919 Forestry Act.

25. See Edward Stebbing, *British Forestry: Its Present Position and Outlook after the War* (London: John Murray, 1916), 15.

26. C. C. Robertson, *Some Suggestions as to the Principles of the Scientific Naturalisation of Exotic Forest Trees* (Cape Town, 1910), 1.

27. The poor health of Australia's largest exotic plantations, located in southeast South Australia, was cause for a Royal Commission in 1935 into the health of the forests. See *1935 South Australia Forestry Royal Commission* (Adelaide: Government Printer, 1936).

28. See the authoritative text by William Schlich, *Schlich's Manual of Forestry, Vol. III: Forest Management*, 5th ed. (London: Bradbury, Agnew & Co., 1925), 274.

29. W. E. Hiley, *A Forestry Venture* (London: Faber and Faber, 1964), 105.

30. W. E. Hiley, *Conifers: South African Methods of Cultivation* (London: Faber and Faber, 1959).

31. For a detailed discussion of the debate over Craib's methods, see the proceedings of the 1935 Empire Forestry Conference and E. H. F. Swain's private diary of the 1948 Empire Forestry Conference. State Library of New South Wales ML MSS 2071, E. H. F. Swain, "By the Way: Pepysian Jottings Empire Forestry Conference 1947"; W. H. Guillebaud, "Some Recent Developments in Forest Research," *Forestry (1948)* 22, no. 2: 145–157, 149. Also, see the A. J. O'Connor's preface in I. J. Craib, *Thinning, Pruning, and Management Studies on the Main Exotic Conifers Grown in South Africa* (Pretoria: Government Printer, 1939). Until the publication of Hiley's book on Craib's methods, foresters in Europe had little knowledge of South African methods. See "Reviews," *Irish Forestry* 16, no. 21 (1959): 51.

32. John Phillips, "Ecology the Foundation of Forestry," *Empire Forestry Journal* 10 (1931): 11.

33. For examples of criticisms, see Ray Borne, "The Place of Thinning in Wattle Silviculture and Its Bearing on the Management of Exotic Conifers," *Empire Forestry Journal* 13 (1934), 193–196; W. H. Guillebad, "The Fourth British Empire Forestry Conference, South Africa 1935," *Forestry*, 146.

34. For discussions of Craib's ideas and the acknowledgment of his influence, see *Proceedings of the Ninth International Conference of Agricultural Economists*, 9 (1955) (Menasha, WI: George Banta Pub. Co., 1956), 300; FAO, *Pulp and Paper Prospects in Latin America* (United Nations, 1955), 97; *Proceedings of the Conference on Southern Industrial Forest Management*, Issue 407 (Durham, NC: Duke University School of Forestry, 1960), 47; William Allen Duerr,

Timber!—Problems, Prospects, Policies (Ames: Iowa State University Press, 1972), 115, 118; S. D. Richardson, "Innovation and Prosperity in Forestry?," *New Zealand Journal of Forestry* 20 (1975), 223; H. V. Hinds, "New Zealand's Exotic Forests," *Unalsylva* 17 (1963); S. A. Y. Ormule, *Growth and Yield 35 Years after Commercially Thinning 50-Year-Old Douglas-Fir, Report 21* (Victoria: B.C. Ministry of Forests, FRDA, 1988).

35. Hiley, *Conifers.*

36. Heyns Kotze and Francois Malan, "Further Progress in the Development of Prediction Models for Growth and Wood Quality of Plantation-Grown Sawtimber in South Africa," in *Forest Growth and Timber Quality: Crown Models and Simulation Methods for Sustainable Forest Management: Proceedings of an International Conference*, ed. Dennis P. Dykstra and Robert Monserud (Washington, DC: USFS, 2007), 117.

37. Peter Attiwil and Christopher Weston, "Forest Soils," in *The Forests Handbook, An Overview of Forest Science* ed. Julian Evans (Oxford: Blackwell, 2001), 176–177; Roger Sands, *Forestry in a Global Context* (Oxford: CABI Publishing 2005), 191–195.

38. For the story of Hiley's venture, see his autobiographical book on his experience there. Hiley, *A Forestry Venture*, 184.

39. Ibid., 21.

40. Lawrence Earley, *Looking for Longleaf: The Fall and Rise of an American Forest* (Chapel Hill: University of North Carolina Press, 2004).

41. A. J. Watson and W. J. Cohen, "Pulping of Eucalypts—An Historical Survey," *APPITA* 22, no. 4 (1969): 17–31.

42. R. N. Parker, *Eucalyptus Trials in the Simla Hills* (Calcutta: Government Printer, 1925), 11.

43. Benedict Taylor, "Trees of Gold and Men Made Good? Grand Visions and Early Experiments in Penal Forestry in New South Wales, 1913–1938," *Environment and History* 14, no. 4 (2008): 545–562.

44. Neil Maher, *Nature's New Deal: The Civilian Conservation Corps and the Roots of the American Environmental Movement* (New York: Oxford University Press, 2008), 54.

45. "The Demand for Timber Grows Greater Every Year and the World's Reserves Are Dwindling Rapidly—A World Timber

Shortage," *Western Mail* (Perth, WA: 1885–1954), May 20, 1948: 60, http://nla.gov.au/nla.news-article39086467, accessed March 1, 2012.

46. Sir Archibal Harris, *Timber at War: An Account of the Organisation and Activities of the Timber Control 1939–1945* (London: Ernest Benn, 1965), viii.

47. "History of the Forestry Commission," http://www.forestry.gov.uk/forestry/CMON-4UUM6R, accessed March 2, 2012.

48. FAO, *Report on World Commodity Problems* (Washington: FAO, 1949; repr. New York: Arno Press, 1976), 65.

49. Junichi Iwamoto, "The Development of Japanese Forestry," in *Forestry and the Forest Industry in Japan*, ed. Yoshiya Iwai (Vancouver: University of British Columbia Press, 2003), 8.

50. For the 4 billion ha statistic for 1948, see "Global Forest Resource Assessment," *FAO Forestry Paper 163* (Rome: FAO, 2010), 334.

51. Ibid., 334.

52. For these statistics, see Douglas W. MacCleery, *American Forests: A History of Resilience and Recovery*, 2nd ed. (Durham: Forest History Society, 2002), 33–39.

53. "Key Issues Affecting the Growth and Develoment of the American Tree Farm System," in file The American Tree Farm System 1985, Box 2, American Tree Farm System Records, Forest History Society.

54. Laurence C. Walker, *The Southern Forest: A Chronicle* (Austin: University of Texas Press, 1991), 128.

55. Thomas R. Cox, *The Lumberman's Frontier: Three Centuries of Land Use, Society, and Change in America's Forests* (Corvallis: Oregon State University Press, 2010), 233.

56. H. H. Champan, "Do We Want a 'Pulpwood' Economy for our Southern National Forests?," *American Forests* (1954), 40–42. As cited in Albert Way, *Conserving Southern Longleaf: Herbert Stoddard and the Rise of Ecological Land Management* (Athens: University of Georgia Press, 2011), 186.

57. For a firsthand account, see Bruce Zobel and Jerry Sprague, *A Forestry Revolution: The History of Tree Improvement in the Southern United States* (Durham, NC: Carolina Academic Press, 1993).

58. For the information in this paragraph, see Rowland Burdon and William Libby, *Genetically Modified Forests: From Stone Age to*

Modern Biotechnology (Durham, NC: Forest History Society, 2006), 25–29.

59. Bruce Zobel, Phillip Carman Wakeley, and Jerry R. Sprague, *A Forestry Revolution: The History of Tree Improvement in the Southern United States* (Durham, NC: Forest History Society, 1993), 3–4.

60. E. N. Andrade, *A Cultura dos Eucalyptus* (São Paulo: Brazil de Rothschild & Comp, 1909); *idem, Manual do Plantador de Eucaliptos* (São Paulo: Rothschild & Comp, 1911); E. N. Andrade and O. Vecchi, *Os eucalyptos: sua cultura e exploracao* (São Paulo: Typographia Brazil de Rothschild & Comp, 1918); E. N. Andrade, *Eucalipto* (São Paulo: Chacaras e Quintais, 1939).

61. Warren Dean, *With Broadax and Firebrand: The Destruction of the Brazilian Atlantic Forest* (Berkeley and Los Angeles: University of California Press, 1995), 259.

62. Jaakko Pöyry Oy, "A Case History of the Aracruz Pulp Mill Project in Brazil," in *FAO/Finland Expert Consultation on Appropriate Forest Industries* (Rome: FAO 1986), 342.

63. M. Patricia Marchak, *Logging the Globe* (Montreal: McGill-Queen's University Press, 1995), 303–304.

64. John Dargavel, *Fashioning Australia's Forests* (Melbourne: Oxford University Press, 1995), 231.

65. Rudyard Kipling, "In the Rukh," *MacClures Magazine* (1896), 23.

66. Richard Neville Parker, *Eucalyptus in the Plains of North West India* (Delhi: Government of India Central Publication, 1925), 4–5.

67. Berthold Ribbentrip, *Forestry in British India* (1900; repr. New Delhi: India Publishing Company, 1989), 138.

68. Thomas Latter, "Extract of Report on the Teak Forests in the Tenasserim Provinces," *Journal of the Agricultural and Horticultural Society of India* 6 (1848), 167.

69. S. Eardley-Wilmont, *Report of Forest Administration in British India from the Year 1906–1907* (Calcutta: Government of India, 1908), 14–15.

70. See, for instance, Peter Dauvergne's criticism of sustainable yield in his "Scientific Forestry and Environmental Failures," in *Loggers and Degradation in the Asia-Pacific* (Cambridge: Cambridge University Press, 2001), 33–50.

71. *Bulletin 18, Proceedings of the Duke University Tropical Forestry Symposium, School of Forestry, Duke University, April 21–26, 1965* (Durham: Duke University, School of Forestry, 1965), 27.

72. John Phillips, *The Development of Agriculture and Forestry in the Tropics: Patterns, Problems, and Promise* (London: Faber and Faber, 1961), 131.

73. D. Pandey and Jim Ball, "The Role of Industrial Plantations in Future Global Fibre Supplies," *Unasylva: An International Journal of the Forestry and Food Industries* 49, 2 (1998): 37–43; figures from Jean-Paul Lanly, "Tropical Forest Resources," *FAO Forestry Paper 30* (Rome: FAO, 1982).

74. *World Forest Inventory 1963* (Rome: FAO, 1965).

75. Terry West, "USDA Forest Service Involvement in Post–World War II International Forestry," in *Changing Tropical Forests: Historical Perspectives on Today's Challenges in Central and South America,* ed. Harold Steen and Richard Tucker (Durham, NC: Duke University Press, 1992), 284; for Britain's use of forestry as policy and intelligence, see Gregory Barton, "Environmentalism, Development and British Policy in the Middle East 1945–1965," *Journal of Imperial and Commonwealth History* 38, no. 4 (2010): 653–674.

76. Jack Westoby, "Forest Industries in the Attack on Underdevelopment," *Unasylva: An International Journal of the Forestry and Food Industries* 16, no. 4 (1962): 168–201.

77. Thomas Rundel, *Tropical Forests: Regional Paths of Destruction and Regeneration in the Late Twentieth Century* (New York: Columbia University Press, 2005), 142–143.

78. See discussions of the relationship between state foresters and rural Indians in Gadil and Guha, "Ecological Conflicts and the Environmental Movement in India."

79. Eric Lundqvist, *Extension of Plantation Forestry: Report to the Government of India* (Rome: FAO, 1964), v.

80. Causal factors of deforestation are determined by a meta-analysis of research publications examining the causes of deforestation. See Thomas K. Rudel, Kevin Flesher, Diane Bates, and Sandra Baptista, "Tropical Deforestation Literature: Geographical and Historical Patterns in the Availability of Information and the Analysis of Causes," *Forest Resource Assessment Working Paper* 27 (2000): 12. See also Rundel, *Tropical Forests*, chapter 2.

81. Figures extrapolated from the growth of 8 million hectares of timber plantations from the 1950s to 1980. For statistics on plantations in 1980, see Norman Myers, *Conversion of Tropical Moist Forests: A Report Prepared by Norman Myers for The Committee on Research Priorities in Tropical Biology of the National Research Council* (Washington, DC: National Academy of Sciences, 1980), 34.

82. The results were published in 1982. FAO, *Manual of Forest Inventory with Special Reference to Mixed Tropical Forests*, FAO Paper 27 (Rome: FAO, 1982), 48.

83. FAO, *World Symposium on Man-Made Forests and Their Industrial Importance* (Rome: FAO 1967), 9.

84. Constance McDermott et al., *Global Environmental Forest Policies: An International Comparison* (London: Earthscan, 2010), 162.

85. Robin W. Daughty, *The Eucalyptus: A Natural and Commercial History of the Gum Tree* (Baltimore: Johns Hopkins University Press, 2002), 134.

86. Madhav Gadgil and Ramachandra Guha, *Ecology and Equity: The Use and Abuse of Nature in Contemporary India* (London: Routledge, 1995), 50–51.

87. S. J. George, B. Kumar, and G. R. Rajiv, "Nature of Secondary Succession in the Abandoned Eucalyptus Plantations of Neyyar (Kerala) in Peninsular India," *Journal of Tropical Forest Science* 5 (1993): 372–386, 377.

88. Harold K. Steen, ed., *Plantation Forestry in the Amazon: The Jari Experience* (Durham: Forest History Society, 1997); Jan Laarman and Roger Sedjo, *Global Forests: Issues for Six Billion People* (New York: McGraw Hill, 1992), 154–155.

89. See the interview by John C. Welker in Steen, *Plantation Forestry in the Amazon*, 234–235.

90. Jack Westoby, *The Purpose of Forests: Follies of Development* (Oxford: Blackwell, 1987), 208–209.

91. Jan G. Laarman and Roger A. Sedjo, *Global Forests: Issues for Six Billion People* (New York: McGraw-Hill, 1992), 40–44.

92. See case studies from the seminar report. FAO, *Forestry Paper 7: Forestry for Local Community Development* (Rome: FAO, 1978).

93. *Community Forestry: Ten Years in Review* (Rome: FAO, 1992). See also http://www.fao.org/docrep/u5610e/u5610e04.htm#COMMUNITYpercent20FORESTRY, accessed December 27, 2012. See also FAO, *Forestry Paper 7: Forestry for Local Community Development* (FAO: Rome 1978).

94. Ramachandra Guha, "The Prehistory of Community Forestry in India," *Environmental History* 6, no. 2 (2001): 213–238, 223–224.

95. World Bank, *Forestry Sector Policy Paper* (Washington, DC: World Bank, 1978).

96. FAO Forestry Department, *FAO Forestry Paper 26: Forestry for Rural Development* (Rome: FAO, 1981), 1.

97. Ibid., 2.

98. For a full analysis of global schemes, see C. McDermott, B. Cashore, and P. Kanowski, *Global Environmental Forest Policies* (London: Earthscan Publications Ltd., 2010).

99. See the report of Narinchai Patanapongsa, *Resources and Constraints of Forestry in Thailand: Guidelines for the Establishment of Forestry Extension in the Royal Forest Department, Thailand* (Bangkok: FAO, 1987).

100. In the 1980s, the *Economic and Political Weekly* hosted a number of articles debating the pros and cons of eucalyptus planting. For a sampling of these rich and spirited debates see Mahasveta Devi, "Eucalyptus: Why?," *Economic and Political Weekly* 18, no. 6 (1983): 1379–1381; V. J. Patel, "Rational Approach towards Fuelwood Crisis in Rural India," *Economic and Political Weekly* 20 (1985): 1366–1368; J. Bandyopadhyay and Vandana Shiva, "Eucalyptus in Rainfed Farm Forestry: Prescription for Desertification," *Economic and Political Weekly* 20 (1985): 1687–1688; D. M. Chandrashekhar, B.V. Krishna Murti, and S. R. Ramaswamy, "Social Forestry in Karnataka: An Impact Analysis," *Economic and Political Weekly* 22 (1987): 935–941; and Shyam Sunder and S. Parameswarappa, "Social Forestry and Eucalyptus," *Economic and Political Weekly* 24 (1989): 51–52.

101. For example, see the works of Jayanta Bandyopadhyay.

102. As of 1995, 76.3 percent of plantations in developing nations were located in Asia, the vast majority in India, Indonesia, Thailand, the Philippines, and China. See D. Pandey, "Tropical Forest Plantation Areas: 1995 Output of Project Hardwood Plantations in the Tropics and Sub-tropics," *FAO GCP/INT/628/UK* (1997). Cited in M. I.

Varmola and J. B. Carle, "The Importance of Hardwood Plantations in the Tropics and Subtropics," *International Forestry Review* 4, no. 2 (2002): 110–121.

103. Commonwealth countries, especially Australia and Britain, led research on this issue. For a culmination of this research see Derek Webb et al., *A Guide to Species Selection for Tropical and Sub-Tropical Plantations*, 2nd ed. (Oxford: Commonwealth Forestry Institute, 1984).

104. Norman Myers, *Conversion of Tropical Moist Forests: A Report Prepared by Norman Myers for the Committee on Research Priorities in Tropical Biology of the National Research Council* (Washington, DC: National Academy of Sciences, 1980), 34.

105. Tropical deforestation also became an issue of great public concern in developed countries. See Norman Myers, *The Primary Source: Tropical Forests and Our Future* (New York and London: W. W. Norton, 1984).

106. World Resources Institute, *Tropical Forests: A Call to Action* (World Resources Institute, 1985), 17.

107. Peter Dauvergne, *Shadows in the Forest: Japan and the Politics of Timber in Southeast Asia* (Cambridge, MA: MIT Press, 1997), 165.

108. Ibid., 82.

109. Larry Lohmann, "Freedom to Plant: Indonesia and Thailand in a Globalizing Pulp and Paper Industry," in *Environmental Changes in South-East: People, Politics and Sustainable Development*, ed. Michael Parnwell and Raymond Bryant (London and New York: Routledge, 1996), 23–48.

110. See Richard Carrere, Larry Lohmann and Lawrence Lohmann, *Pulping the South: Industrial Tree Plantations and the World Paper Economy* (London: Zed Books, 1996); Winfridus Overbeek, Markus Kröger, and Julien-François Gerber, "An Overview of Industrial Tree Plantations in the Global South," *EJOLT Report No. 3* (2012).

3 Native Forests

1. William Lines, *Taming the Great South Land: A History of the Conquest of Nature in Australia* (Berkeley: University of California Press, 1991), 259.

2. Raymond Williams, *Keywords: A Vocabulary of Culture and Society*, 2nd ed. (Oxford: Oxford University Press, 1983), 218. See William Cronon, "The Trouble with Wilderness; or, Getting Back to the Wrong Nature," in *Uncommon Ground: Rethinking the Human Place in Nature*, ed. William Cronon (New York: W. W. Norton & Co., 1995), 69–90.

3. Nancy Langston, *Forest Dreams, Forest Nightmares: The Paradox of Old Growth in the Inland West* (Seattle: University of Washington Press, 1995), 148.

4. David Lindenmayer, ed., *Forest Pattern and Ecological Process: A Synthesis of 25 Years of Research* (Collingwood: CSIRO, 2009), 217.

5. Hirt, *A Conspiracy of Optimism*, xxxii–xxxvii.

6. Theodore Porter, *Trust in Numbers: The Pursuit of Objectivity in Science and Public Life* (Princeton: Princeton University Press, 1995), 115.

7. Ibid., 114–117.

8. *The Secretary of State for India to the Governor General of India in Council, 14 Sept 1866, in A Selection of Despatches and their Enclosures to and from the Secretary of State for India in Council on Forest Conservancy in India, showing the Measures which have been adopted, and the Operations which are going on in the several Presidencies and Lieutenant Governorships, beginning with the Despatch from the Governor General in Council of the 21st day of May 1862 to the present time Vol. 1* (London: Parliament of Great Britain, 1871).

9. Tamara Whited, *Forest and Peasant Politics in Modern France* (New Haven: Yale University Press, 2000), 31–34.

10. *A Selection of Despatches*.

11. Foresters and hydrologists differed into the mid-twentieth century. See W. G. Hoyt and W. B. Langhein, *Floods* (Princeton: Princeton University Press, 1955), 155.

12. See Michael Imort, "A Sylvan People: Wilhelmine Forestry and the Forest as a Symbol of Germandom," in *Germany's Nature: Cultural Landscapes and Environmental History*, ed. Thomas M. Lekan and Thomas Zeller (Piscataway, NJ: Rutgers University Press, 2005), 62.

13. Hölzl, "Historicizing Sustainability," 449–450.

14. Von Salisch, *Forest Aesthetics*.

15. O. Ciancio and S. Nocentini, "Forest Management from Positivism to the Culture of Complexity," in *Forest History International Studies on Socioeconomic and Forest Ecosystem Change*, ed. M. Agnoletti and S. Anderson (Oxon: CABI Publishing, 2000), 49.

16. A. K. Cajander, "Über walden typen," *Acta Forestalia Fennica*, 1 (1909).

17. Michael Imort, "'Eternal Forest-Eternal Volk': The Rhetoric and Reality of National Socialist Forest Policy," in *How Green Were the Nazis?: Nature, Environment, and Nation in the Third Reich*, ed. Franz-Josef Brüggemeier, Mark Cioc, and Thomas Zeller (Athens: Ohio University Press, 2005), 46–48.

18. Frank Uekötter, *The Green and the Brown: A History of Conservation in Nazi Germany* (Cambridge: Cambridge University Press, 2006), 69–71.

19. Imort, *Eternal Forest—Eternal Volk*, 49.

20. Frank Uekötter, *The Greenest Nation?: A New History of German Environmentalism* (Cambridge, MA: MIT Press, 2014), 51.

21. See Aldo Leopold, "Deer and Dauerwald in Germany: I. History," *Journal of Forestry* 34, no. 4 (1936): 366–375; Aldo Leopold, "Deer and Dauerwald in Germany: II. Ecology and Policy," *Journal of Forestry* 34, no. 5 (1936): 460–466.

22. Harold Steen, *The U.S. Forest Service: A Centennial History* (Durham, NC: Forest History Society in association with University of Washington Press, 2004), 154–157.

23. See L. F. Cook, memorandum for the Chief Forester, Reply to Dr. Murie's Report on the Glacier National Park Cleanup Project, 28 August 1935, Entry 34, Record Group 79, Records of the National Park Service, National Archives. As cited in Richard Sellars, "Science and Natural Resource Management in the National Parks Service, 1929–40," in *Forest and Wildlife Science in America: A History*, ed. Harold K. Steen (Durham, NC: Forest History Society, 1999), 115.

24. E. A. Grewswel, "The Constructive Properties of Fire in Chir (*Pinus longifolia*) Forests," *Indian Forester* 52 (1926): 502–505. As cited in *Proceedings of the First Tall Timbers Fire Ecology Conference* 1 (1962).

25. J. Keet, *Fourth British Empire Forestry Conference Proceedings* (Pretoria: Government Printer, 1935), 98.

26. "Forestry Research Essential: General Smuts Stresses Need for Scientific Development," *Rand Daily Mail*, 9 September 1935.

27. Lance van Sittert, "Making the Cape Floral Kingdom: The discovery and Defense of Indigenous Flora at the Cape ca. 1890-1939," *Landscape Research*, 28, no. 1 (2003): 113–129.

28. Keet, *Fourth British Empire Forestry Conference Proceedings*, 25.

29. Ibid., 25.

30. C. W. Wicht, *Forestry and Water Supplies in South Africa* (Pretoria: Government Printer, 1949).

31. See also C. L. Wicht, 'Summary of Forests and Evapotranspiration Session,' in *International Symposium on Forest Hydrology, Penn State University* 11, no. 1 (1967).

32. Brett Bennett and Frederick Kruger, *Forestry and Water Conservation in South Africa: History, Science, Policy* (Canberra: ANU Press, 2015).

33. Emily Brock, "The Challenge of Reforestation: Ecological Experiments in the Douglas Fir Forests 1920 to 1940," *Environmental History* 9, no. 1 (2004): 57–79, 66–67.

34. See, for instance, the findings of the Royal Commission in British Columbia. Peter H. Pearse, *Timber Rights and Forest Policy in British Columbia: Report of the Royal Commission on Forest Resources* (Victoria: The Commission, 1976).

35. Thomas Dunlop, *Nature and the English Diaspora: Environment and History in the United States, Canada, Australia, and New Zealand, No. 17* (New York: Cambridge University Press, 1999).

36. Amelia R. Pry, ed., *Douglas Fir Research in the Pacific Northwest, 1920–1956, an Interview with Leo A. Isaac* (Oral History, Bancroft Library, University of California at Berkeley, 1967).

37. Brock, *The Challenge of Reforestation*, 66–67.

38. Griffiths, *Forests of Ash*, 163–165.

39. A. D. Johnston, "Management of West Coast Beech Forests," *New Zealand Journal of Forestry* 17, no. 2 (1972): 180–188, 180, 183.

40. H. L. Harper and J. White, "The Demography of Plants," *Annual Review of Ecology and Systematics* 5 (1974): 419–463.

41. Frederick H. Swanson, *The Bitterroot and Mr. Brandborg: Clearcutting and the Struggle for Sustainable Forestry in the Northern Rockies* (Salt Lake City: University of Utah Press, 2011), 3.

42. *Newsweek*, June 1952, cited in Doug MacCleery, "Re-inventing the United States Forest Service: Evolution from Custodial Management to Protection Forestry, to Ecosystem Management," in *Re-inventing Forestry Agencies Experiences of institutional restructuring in Asia and the Pacific* (Bangkok: FAO, 2008).

43. Westoby, *The Purpose of Forests*, 122.

44. Susan Schrepfer, "Conflict in Preservation: The Sierra Club, Save the Redwoods League, and Redwood National Park," *Journal of Forest History* 24 (1980): 60–76.

45. Emanuel Fritz, "The Development of Industrial Farming in California, The Colonel William B. Greely Lectures," *in Industrial Forestry* (Seattle: University of Washington College of Forestry, 1960), 13.

46. Inspection Records, 1942–43, 1970, Box 2, American Tree Farm System Records, Forest History Society.

47. Stephanie S. Pincetl, *Transforming California: A Political History of Land Use and Development* (Baltimore: Johns Hopkins University Press, 1999), 165.

48. Emanuel Fritz to Samuel T. Dana, January 12, 1965, American Forestry Association Archives, Box 65, Forest History Society.

49. David Brower, *For Earth's Sake: The Life and Times of David Brower* (Salt Lake City: Peregrine Smith Books, 1990), 362.

50. Jared Farmer, *Trees in Paradise: A California History* (New York: W. W. Norton, 2013), 82.

51. Kathryn Newfont, *Blue Ridge Commons: Environmental Activism and Forest History in Western North Carolina* (Athens: University of Georgia Press), 134–141.

52. Swanson, *The Bitterroot and Mr. Brandborg*.

53. Newfont, *Blue Ridge Commons*.

54. Ibid., 88.

55. John Dargavel, *Fashioning Australia's Forests* (Melbourne: Oxford University Press, 1995), 87.

56. Richard Routley and Val Routley, *The Fight for the Forests: The Takeover of Australian Forests for Pines, Wood Chips and*

Intensive Forestry, 3rd ed. (Canberra: Research School of Social Sciences, 1975), 189.

57. Ibid., 189.

58. Ibid., 168.

59. Ibid., 168–196.

60. New Zealand Forest Service, *Utilization of South Island Beech Forests* (Wellington: A. R. Shearer, government printer, 1971).

61. Rob Nixon, *Slow Violence and the Environmentalism of the Poor* (Cambridge, MA: Harvard University Press, 2011).

62. *Commonwealth v. Tasmania.*

63. Graham Richardson in *Quarterly Essay: Beautiful Lies: Population and Environment in Australia*, 9 (2003): 79.

64. See the influential report by Colin Griffiths, *A Study of the Conservation Significant of the West Tropics of North-East Queensland: A Report to the Australian Heritage Commission* (Bardon: The Society, 1984).

65. W. J. Hurdich, "Problems of Public Forestry and the Socio-Economic Implications of Privatisation," *Oxford Forestry Institute Occasional Papers*, no. 42 (Oxford: Oxford Forestry Institute, 1993), 3.

66. As cited in Douglas Bevington,*The Rebirth of Environmentalism: Grassroots Activism from the Spotted Owl to the Polar Bear* (Washington, DC: Island Press, 2009), 121.

67. See David Brooks and Gordon Grant, *New Perspectives in Forest Management: Background, Science Issues, and Research Agenda, Res. Pap. PNW-RP-456* (Portland, OR: U.S. Department of Agriculture, 1992), 7.

68. Harold Steen, ed., *An Interview with F. Dale Robertson* (Durham, NC: Forest History Society, 1999), 86–88.

69. Dale Robertson, *USFS Chief's Report 1990* (Washington, DC: USFS, 1990), 17.

70. Jack Ward Thomas et al. "The Northwest Forest Plan: Origins, Components, Implementation Experience, and Suggestions for Change," *Conservation Biology* 20, no. 2 (2006): 277–287.

71. Robert Nelson, "The Religion of Forestry: Scientific Management," *Journal of Forestry* 97, no. 11 (1999): 4–8, 7.

72. Brett Bennett, "El Dorado of Forestry: The Eucalyptus in India, South Africa, and Thailand, 1850–2000," *International Review of Social History* 55 (2010): 27–50, 46–47.

73. Tim Forsyth and Andrew Walker, *Forest Guardians, Forest Destroyers: the Politics of Environmental Knowledge in Northern Thailand* (Seattle: University of Washington Press, 2008).

74. Runsheng Yin et al., "China's Ecological Rehabilitation: The Unprecedented Efforts and Dramatic Impacts of Reforestation and Slope Protection in Western China," *China Environment Series7* (Washington, DC: Woodrow Wilson International Center for Scholars, 2005), 20–21.

75. See Dauvergne and Lister, *Timber*.

76. Environmental Investigations Agency, *First Class Crisis: China's Criminal and Unsustainable Intervention in Mozambique's Mimbo Forests* (London: EIA, 2014).

77. Ministry of Forests, Mines and Lands, *State of British Columbia's Forests, Third Edition 2010* (Victoria: Forest Practices and Investment Branch, 2010).

4 Toward a Twenty-First-Century Consensus

1. Saskia Ozinga and Hannah Mowat, "Strategies to Prevent Illegal Logging," in *A Handbook of Globalisation and Environmental Policy, Second Edition: National Government Interventions in a Global Arena*, ed. Frank Wijen et al. (Cheltenham: Edward Elgar, 2012), 442.

2. David J. Brooks et al., *Economic and Environmental Effects of Accelerated Tariff Liberalization in the Forest Products Sector* (Portland, OR: U.S. Forest Service, 2001), 2.

3. See, for instance, Susanna Myllylä and Tuomo Takala, "Leaking Legitimacies: The Finnish Forest Sector's Entanglement in the Land Conflicts of Atlantic Coastal Brazil," *Social Responsibility Journal* 7, no. 1 (2011): 42–60.

4. FAO, *State of the World's Forests Report*, 93.

5. International Tropical Timber Organization, "Encouraging Industrial Forest Plantations in the Tropics: Report of a Global Study," *Technical Series* 33 (2009): 11.

6. Dauvergne and Lister, *Timber*, 19.

7. Claudia Stickler, "The Potential Ecological Costs and Co-benefits of REDD: A Critical Review and Case Study from the Amazon Region," *Global Change Biology* 15, no. 12 (2009): 2803–2824.

8. David Richardson et al., "Invasive Agroforestry Trees: Problems and Solutions," *Agroforestry and Biodiversity in Tropical Landsacapes*, ed.Götz Schroth et al. (Washington, DC: Island Press, 2004): 371–396.

9. Judith Adjani, "Election 2007: Ending the Forest Wars," *Australian Review of Public Affairs*, September 2007, http://www.australianreview.net/digest/2007/election/ajani.html, accessed July 7, 2014. See also Judith Ajani, *The Forest Wars* (Melbourne: Melbourne University Press, 2007).

10. Peichang Zhang et al., "China's Forest Policy for the 21st Century," *Science* 288, no. 5474 (2000): 2135–2136.

11. Richard Hobbs et al., "Intervention Ecology: Applying Ecological Science in the Twenty-First Century," *BioScience* 61, no. 6 (2011): 442–450.

12. James E. Watson et al., "The Performance and Potential of Protected Areas," *Nature* 515 (2014): 67–73.

13. William Laurance, "The Perils of Payoff: Corruption as a Threat to Global Biodiversity," *Trends in Ecology and Evolution* 19, no. 8 (2004): 399–401.

14. Tom Clements, "Reduced Expectations: The Political and Institutional Challenges of REDD+," *Oryx: International Journal of Conservation* 44, no. 3 (2010): 309–310.

15. J. R. Vincent, "The Tropical Timber Trade and Sustainable Development," *Science* 256, no. 5064 (1992): 1651–1655.

16. Alain Paquette and Christian Messier, "The Role of Plantations in Managing the World's Forests in the Anthropocene," *Frontiers in Ecology and the Environment* 8, no. 1 (2010): 27–34, 29.

17. Walter Baber and Robert Bartlett, *Deliberative Environmental Politics: Democracy and Economic Rationality* (Cambridge, MA: MIT Press, 2005).

18. Susan Charnley and Melissa R. Poe, "Community Forestry in Theory and Practice: Where Are We Now?," *Annual Review of Anthropology* 36 (2007) 301–336.

19. Andrew Matthews, *Instituting Nature: Authority, Expertise, and Power in Mexican Forests* (Cambridge, MA: MIT Press, 2011).

20. See, for instance, Jacki Schirmer, "Plantations and Social Conflict: Exploring the Differences between Small-Scale and Large-Scale Plantation Forestry," *Small-Scale Forestry* 6, no. 1 (2007): 19–33. The examples come from developed countries but are applicable to developing countries.

21. Benjamin William Cashore, Graeme Auld, and Deanna Newsom, *Governing through Markets: Forest Certification and the Emergence of Non-state Authority* (New Haven: Yale University Press, 2004).

22. Peter Kanowski et al., "House of Representatives Standing Committee on Agriculture, Resources, Fisheries and Forestry Inquiry into the Australian Forestry Industry," no date [2000s].

23. David Gaveau et al., "Reconciling Forest Conservation and Logging in Indonesian Borneo," *PLoS One* 8, no. 8 (2013): e69887, doi:10.1371/journal.pone.0069887.

Selected Bibliography

Albritton-Jonsson, Fredrik. *Enlightenment's Frontier: The Scottish Highlands and the Origins of Environmentalism*. New Haven: Yale University Press, 2013.

Appuhn, Karl. *A Forest on the Sea: Environmental Expertise in Renaissance Venice*. Baltimore: Johns Hopkins University Press, 2009.

Baber, Walter, and Robert Bartlett. *Deliberative Environmental Politics: Democracy and Economic Rationality*. Cambridge, MA: MIT Press, 2005.

Barton, Gregory. *Empire Forestry and the Origins of Environmentalism*. Cambridge: Cambridge University Press, 2002.

Beattie, James. *Empire and Environmental Anxiety: Health, Science, Art and Conservation in South Asia and Australasia, 1800–1920*. Basingstoke: Palgrave Macmillan, 2011.

Bellich, James. *Replenishing the Earth: The Settler Revolution and the Rise of the Anglo-World, 1783–1939*. Oxford: Oxford University Press, 2009.

Bevington, Douglas. *The Rebirth of Environmentalism: Grassroots Activism from the Spotted Owl to the Polar Bear*. Washington, DC: Island Press, 2009.

Beinart, William, and Lotte Hughes. *Environment and Empire*. Oxford: Oxford University Press, 2007.

Brooks, David J., Joseph A. Ferrante, Jennifer Haverkamp, Ian Bowles, William Lange, and David Darr. *Economic and Environmental Effects of Accelerated Tariff Liberalization in the Forest Products Sector*. Portland, OR: U.S. Forest Service, 2001.

Carrere, Richard, Larry Lohmann, and Lawrence Lohmann. *Pulping the South: Industrial Tree Plantations and the World Paper Economy*. Atlantic Highlands, NJ: Zed Books, 1996.

Cashore, Benjamin William, Graeme Auld, and Deanna Newsom. *Governing Through Markets: Forest Certification and the Emergence of Non-state Authority*. New Haven: Yale University Press, 2004.

Cohen, Shaul. *Planting Nature: Trees and the Manipulation of Environmental Stewardship in America*. Berkeley: University of California Press, 2004.

Cox, Thomas R. *The Lumberman's Frontier: Three Centuries of Land Use, Society, and Change in America's Forests*. Corvallis: Oregon State University Press, 2010.

Cronon, William. "The Trouble with Wilderness; or, Getting Back to the Wrong Nature." In *Uncommon Ground: Rethinking the Human Place in Nature*, ed. William Cronon, 69–90. New York: W. W. Norton & Co, 1995.

Dargavel, John, and Elisabeth Johann. *Science and Hope: A Forest History*., Cambridge: White Horse Press, 2013.

Dargavel, John. *Fashioning Australia's Forests*. Melbourne: Oxford University Press, 1995.

Dauvergne, Peter. *Shadows in the Forest: Japan and the Politics of Timber in Southeast Asia*. Cambridge, MA: MIT Press, 1997.

Daughty, Robin W. *The Eucalyptus: A Natural and Commercial History of the Gum Tree*. Baltimore: Johns Hopkins University Press, 2002.

Dauvergne, Peter. *Loggers and Degradation in the Asia-Pacific*. Cambridge: Cambridge University Press, 2001.

Davis, Diana. *Resurrecting the Granary of Rome: Environmental History and French Colonial Expansion in North Africa*. Athens: Ohio University Press, 2007.

Dean, Warren. *With Broadax and Firebrand: The Destruction of the Brazilian Atlantic Forest*. Berkeley, Los Angeles: University of California Press, 1995.

Earley, Lawrence. *Looking for Longleaf: The Fall and Rise of an American Forest*. Chapel Hill: University of North Carolina Press, 2004.

Evans, Julian. *Plantation Forestry in the Tropics*. 2nd ed. Oxford: Oxford University Press, 1992.

Farmer, Jared. *Trees in Paradise: A California History*. New York: W. W. Norton, 2013.

Flemming, James. *Historical Perspectives on Climate Change*. New York: Oxford University Press, 1998.

Forsyth, Tim, and Andrew Walker. *Forest Guardians, Forest Destroyers: the Politics of Environmental Knowledge in Northern Thailand*. Seattle: University of Washington Press, 2008.

Franklin, Jerry. "Challenges to Temperate Forest Stewardship—Focusing on the Future." In *Towards Forest Sustainability*, ed. David Lindenmayer and Jerry Franklin, 1–13. Collingwood: CSIRO Publishing, 2003.

Gadgil, Madhav, and Ramachandra Guha. *This Fissured Land: An Ecological History of India*. New Delhi: Oxford University Press, 1992.

Gadgil, Madhav, and Ramachandra Guha. *Ecology and Equity: The Use and Abuse of Nature in Contemporary India*. London: Routledge, 1995.

Gissibl, Bernhard, Sabine Höhler, and Patrick Kupper, eds. *Civilizing Nature: National Parks in Global Historical Perspective*. New York: Berghan Books, 2012.

Griffiths, Tom. *Forests of Ash: An Environmental History*. Melbourne: Cambridge University Press, 2001.

Grove, Richard. *Green Imperialism: Colonial Expansion, Tropical Island Edens and the Origins of Environmentalism, 1600–1860*. Cambridge: Cambridge University Press, 1995.

Guha, Ramachandra. *The Unquiet Woods: Ecological Change and Peasant Resistance in the Himalaya*. 2nd ed. Oxford: Oxford University Press, 2000.

Hopkins, A. G., ed. *Globalization in World History*. London: Pimlico, 2002.

Hopkins, A. G., ed. *Global History: Interactions between the Universal and the Local*. Basingstoke: Palgrave Macmillan, 2006.

Krech, Shepard. *The Ecological Indian: Myth and Imagination*. New York: W. W. Norton, 2000.

Langston, Nancy. *Forest Dreams, Forest Nightmares: The Paradox of Old Growth in the Inland West*. Seattle: University of Washington Press, 1995.

Lekan, Thomas M., and Thomas Zeller, eds. *Germany's Nature: Cultural Landscapes and Environmental History*. Piscataway, NJ: Rutgers University Press, 2005.

Lester, A., and D. Lambert, eds. *Colonial Lives across the British Empire: Imperial Careering in the Long Nineteenth Century*. Cambridge: Cambridge University Press, 2006.

Lindenmayer, David. *Forest Pattern and Ecological Process: A Synthesis of 25 Years of Research*. Collingwood: CSIRO Publishing, 2009.

Lines, William. *Taming the Great South Land: A History of the Conquest of Nature in Australia*. Berkeley: University of California Press, 1991.

Livingstone, D. N. *Putting Science in Its Place: Geographies of Scientific Knowledge*. Chicago: University of Chicago Press, 2003.

Lowood, Henry E. "Emergence of Scientific Forestry Management in Germany." In *The Quantifying Spirit in the Eighteenth Century*, ed. Tore Frängsmyr, J. L. Heilbron, and Robin E. Rider, 315–342. Berkeley: University of California Press, 1990.

MacCleery, Douglas W. *American Forests: A History of Resilience and Recovery*. 2nd ed. Durham, NC: Forest History Society, 2002.

Maher, Neil. *Nature's New Deal: The Civilian Conservation Corps and the Roots of the American Environmental Movement*. New York: Oxford University Press, 2008.

Marchak, M. Patricia. *Logging the Globe*. Montreal: McGill-Queens Press, 1995.

Matthews, Andrew. *Instituting Nature: Authority, Expertise, and Power in Mexican Forests*. Cambridge, MA: MIT Press, 2011.

Miller, Char. *Seeking the Greatest Good: The Conservation Legacy of Gifford Pinchot*. Pittsburgh: University of Pittsburgh Press, 2013.

Myers, Norman. *The Primary Source: Tropical Forests and Our Future*. New York: W. W. Norton, 1984.

Newfont, Kathryn. *Blue Ridge Commons: Environmental Activism and Forest History in Western North Carolina*. Athens: University of Georgia Press, 2012.

Porter, Theodore. *Trust in Numbers: The Pursuit of Objectivity in Science and Public Life*. Princeton: Princeton University Press, 1994.

Rajan, Ravi. *Modernizing Nature: Forestry and Imperial Eco-Development 1800–1950*. Oxford: Oxford University Press, 2006.

Richards, John. *The Unending Frontier: an Environmental History of the Early Modern World*. Berkeley: University of California Press, 2005.

Richardson, David M., Pierre Binggeli, and Gotz Schroth. "Invasive Agroforestry Trees: Problems and Solutions." In *Agroforestry and Biodiversity in Tropical Landsacapes*, ed. Götz Schroth, 371–396. Washington, DC: Island Press, 2004.

Rowland, Burdon, and William Libby. *Genetically Modified Forests: From Stone Age to Modern Biotechnology*. 2006. Durham, NC: Forest History Society.

Rundel, Thomas. *Tropical Forests: Regional Paths of Destruction and Regeneration in the Late Twentieth Century*. New York: Columbia University Press, 2005.

Sands, Roger. *Forestry in a Global Context*. Oxford: CABI Publishing, 2005.

Scott, James. *Seeing Like a State: How Certain Schemes to Improve the Human Condition Have Failed*. New Haven: Yale University Press, 1999.

Sivaramakrishnan, K. *Modern Forests: Statemaking and Environmental Change in Eastern India*. Palo Alto, CA: Stanford University Press, 1999.

Steen, Harold. *The U.S. Forest Service: A Centennial History*. Seattle: University of Washington Press, 2005.

Sunseri, Thaddeus Raymond. *Wielding the Ax: State Forestry and Social Conflict in Tanzania, 1820–2000*. Athens: Ohio University Press, 2009.

Swanson, Frederick H. *The Bitterroot and Mr. Brandborg: Clearcutting and the Struggle for Sustainable Forestry in the Northern Rockies*. Salt Lake City: University of Utah Press, 2011.

Totman, Conrad. *Japan: An Environmental History*. London: I. B. Taurus, 2014.

Totman, Conrad. *The Green Archipelago: Forestry in Preindustrial Japan*. Berkeley: University of California Press, 1989.

Tropp, Jacob. *Natures of Colonial Change: Environmental Relations in the Making of the Transkei*. Athens: Ohio University Press, 2006.

Tyrell, Ian. *Crisis of the Wasteful Nation: Empire and Conservation in Theodore Roosevelt's America*. Chicago: University of Chicago Press, 2015.

Uekötter, Frank. *The Green and the Brown: A History of Conservation in Nazi Germany*. Cambridge: Cambridge University Press, 2006.

Uekötter, Frank. *The Greenest Nation? A New History of German Environmentalism*. Cambridge, MA: MIT Press, 2014.

Wall, Derek. *The Commons in History: Culture, Conflict and Ecology*. Cambridge, MA: MIT Press, 2014.

Way, Albert. *Conserving Southern Longleaf: Herbert Stoddard and the Rise of Ecological Land Management*. Athens: University of Georgia Press, 2011.

Westoby, Jack. *The Purpose of Forests: Follies of Development*. Oxford: Blackwell, 1987.

Whited, Tamara. *Forest and Peasant Politics in Modern France*. New Haven: Yale University Press, 2000.

Williams, Michael. *Deforesting the Earth: From Prehistory to Global Crisis*. Chicago: University of Chicago Press, 2003.

Worster, Donald. *The Wealth of Nature: Environmental History and the Ecological Imagination*. New York: Oxford University Press, 1993.

Zobel, Bruce, and Jerry Sprague. *A Forestry Revolution: The History of Tree Improvement in the Southern United States*. Durham: Carolina Academic Press, 1993.

Index